KIDS, PARENTS & TECHNOLOGY:

An Instruction Guide for Young Families

Second Edition

Dr. Eitan D. Schwarz

MyDigitalFamily

Copyright ©2009 Eitan Schwarz. All rights reserved. (www.mydigitalfamily.org.)

Design consultation by Jessica Schwarz.

ISBN 978-0-557-19482-7

DEDICATED WITH DEEPEST RESPECT

TO THE DEVOTED FOUNDERS, STAFF, AND SUPPORTERS

OF CHICAGO'S MATCHLESS

FAMILY FOCUS

BRILLIANTLY PIONEERING ITS GROUND-BREAKING MISSION:

"TO PROMOTE THE WELL-BEING OF CHILDREN FROM BIRTH

BY SUPPORTING AND STRENGTHENING THEIR FAMILIES

IN AND WITH THEIR COMMUNITIES."

www.family-focus.org

We rejoice in the children who bless our homes, whose eager minds and hearts are the promise of tomorrow.

Preface

Parents—you need help. Instruction manuals and guides are commonly boxed together with brand-new technological devices, such as cameras, computers, or software, but none actually tells you how to use the device for your family's well-being. Now you finally have in your hands the instruction manual you've been asking for that will guide you through not only using these devices, but using them well for your family's benefit.

This book provides a systematic child- and family-centered, systematic approach for parents (and therapists, pediatricians, educators, etc.) consisting of clear, practical, and positive guidelines to structure kids' media experiences. The book is peppered with illustrative vignettes and therapy encounters from my practice utilizing these methods.

The Setup and Quick Start Guide will take you through the basics of this new method that will improve your children's interactive media activities and enable you to take full advantage of your family computer and other digital devices.

The comprehensive *User's Guide* will provide a thorough understanding of the needs of children and how their interactions with technology are impacting their lives now and will do so even more in the future.

Food for Thought examines specific issues in more depth.

Parenting Basics, scattered throughout, are insights collected during my years of practice which will give you useful ideas to apply to other areas of your family life and relationships with your children.

You will learn to prepare children for a lifetime of encounters with interactive media. A good portion of the book is devoted to the future because there will surely be more challenges and opportunities than there have been so far. Looking down the road even further, no matter what new technologies they come across, the good habits and discipline you teach your kids at home now will endure throughout their lifetimes.

To be entirely frank, I must own up to several other agendas: To coach parents to teach their children positive habits to serve them in the future, generally. To boost the sensitivity and effectiveness of families with solid and credible information, I show how the best practitioners of child and adolescent psychiatry actually think and work daily. Children and families deserve our best evidence-based thinking as we go forward into a technology- and media-saturated world.

Acknowledgements: I want to thank those I hold closest—my dear family and fabulous friends. You have supported me in this project with your patience, interest, and love. I thank you here, and even more, in my heart.

This is my first chance to affirm publically my gratitude for the privilege of giving care to folks in the same community for almost forty years. I have been truly honored to be your doctor as you have struggled with heavy and painful burdens.

I remember and thank each one of you for making me a wiser physician and kinder person. I thank especially Nathan, Sharon, Henry, Liz, David, Lillian, Josh, Jeremy, and others who taught me and gave their invaluable and resourceful insights. I have always liked learning from kids, but from you guys especially. I hope that you all appreciate that what you taught me could reach many other kids today and in years to come.

This book is the culmination of a life's work. I am grateful and fortunate to have had great teachers: Virginia Rasmussen, who inspired me to write as a youngster in Plainfield, NJ. At Johns Hopkins Medical School, I was privileged to be accepted into a caring academic community. I learned what excellence in medicine really means from Alex Haller, MD, Henry Wagner, MD, and Phillip Tumulty, MD, who showed me how to be a doctor; Leigh Thompson, MD, PhD, who taught me medical research; Jerome Frank, MD, PhD, whose gentleness and wisdom have inspired me throughout my lifetime; and Nahum Spinner, MD, and Seymore Perlin, MD, who led me patiently to their insights. My talented teachers at the University of Chicago Hospitals were William Offenkrantz, MD, Sherman Feinstein, MD, Jarl Dyrud, MD, Israel Goldiamond, PhD, Edward Senay, MD, Edward Stein, MD, Phillip Holtzman, PhD, Nathaniel Apter, MD, Eberhard Uhlenhuth, MD, Harry Trossman, MD, Daniel X. Freedman MD, Charles Kramer, MD, Leonard Elkund, MD, Bernard Rubin, MD, Heinz Kohut, MD, and David Klass, MD. In the community and at Evanston Hospital, I thank Henry Fineberg, MD, Robert Gluckman, MD, Ira Sloan, MD, Ron Rozensky, PhD, and Derek Miller, MD.

In writing this book, I have been fortunate to receive the support of my family, and the kindness, talents, and generosity of many wonderful and capable professionals. I am glad to give them all the respect they deserve and loads of thanks. Among them are especially Drs. Joe Siegler, Dennis Grygotis, Stephanie Wright, Alvin Rosenfeld, Marcia Leiken, and Connie McLoughlin; and especially Janet Haney, Pierre Lehu, Nancy Rosenfeld, and all the others who have shown me the ways of the publishing and marketing worlds. The folks at Lulu.com continued to be supportive and extremely helpful. Other special friends have included Carolyn Schwarz, Kevin Limbeck, David Sullivan, and George Cyrus. I would also like to recognize the scholars, thinkers, and colleagues whose pioneering and advocacy work on behalf of youth interacting with media has given me the insights and joy of saying "aha!" so many times as I researched this book.

Preface

A Note About the Second Edition: Self-publishing is a fortunate adventure that happily brings more opportunities than hazards. Itself a rapidly evolving area, self-publishing has made it possible to quickly bring out this transformed version. I hope the readers like it, and, ultimately, kids and families benefit from it.

While the First Edition contained pretty much all of my basic innovations, it was, after all, meant to be an instruction manual, a bit rough around the edges. With improved appearance, editing, and formatting, the manual has blossomed into a more refined book.

I had the great fortune to work with superb professionals: A fabulous editor in the person of Laura Zuckerman, to whom I am enormously grateful, meticulously manicured the text. Beth Bennett has been a wonderful advisor. Alissa Schwarz made this effort more of a coherent brand. And the talented Jessica Schwarz has upgraded our ever-evolving online presence, integrated closely with the book.

I was able to make changes based on rapid developments in technology and my own increased immersion in it. For example, we added terms like 'smartphone' and 'social media' and eliminated terms like 'surfing', and 'the PC'. We dropped the label 'Media Diet' because it has a negative and restrictive connotation, and we are emphasizing for parents the commitment to plan the positive and constructive use of technology. We also simplified some of our evolving terminology, for example changing 'Family Media Diet' to 'Media Plan', and the names of some of the forms. In addition, we added instructions for creating the Media Portal Page.

TABLE OF CONTENTS

Introduction: Only You Can Give Your Kids This Great Gift............................ 1

PART 1: SETUP AND QUICK START GUIDE 8

    Chapter 1: An Extreme Makeover—Appreciate That You Are Wiring Your Children's Brains 10

    Chapter 2: The Media Plan—Harvesting Growth Opportunities from Media 20

    Chapter 3: Setup and Installation of the Media Plan 34

    Chapter 4: Keeping Your Baby's Brain "Green" (Birth to Two) 50

    Chapter 5: Learning What's Real (Two to Five) 62

    Chapter 6: Collaborating with Your Kids (Five to Eight) 68

Part 2: USER'S GUIDE 86

    Chapter 7: More on Values Education 88

    Chapter 8: Leading Your Family 98

    Chapter 9: Safety and Other Challenges 112

    Chapter 10: Fancy Menus and Special Plans 128

Part 3: FOOD FOR THOUGHT 144

    Chapter 11: Technology and Play: A Different Game................. 146

    Chapter 12: The Future I: Interactive Media Play Therapy and Other Clinical Applications 158

    Chapter 13: The Future II: FITGOALS™ 182

    Chapter 14: The Future III: A Nonviolent Video game................ 190

Index.. 210

FOR END NOTES, PLEASE GO TO WWW.MYDIGITALFAMILY.ORG

Introduction: Only You Can Give Your Kids This Great Gift

Question: Dr. Schwarz, how did you come to write this book?

Answer: Actually, the basic idea came directly from how good people tried to make their lives better. Where there is a will and good faith, there is a way—there is no end to people's inventiveness. This is what actually happened:

Some years ago, eleven-year-old Jason's mother called me for an appointment, frustrated that Jason was out of control. He was spending entirely too much time online, and his schoolwork and social life were suffering. When not online, Jason played video games. None of the self-help books Mom had tried helped. Jason's online roaming had brought a virus into the computer, and Mom, begrudgingly, had to buy a new one. That was the last straw.

Jason was a normal kid from a good home, but expectedly, Mom and Jason had been having difficulty since Dad died a year before. With much concentrated effort, they used therapy to make their way together through painful grieving and difficult rebuilding. Mom came to realize that she had lost her footing as a parent. Jason had turned to interactive media electronic devices for a desperate illusion of premature self-sufficiency and to reduce the overwhelming emotional turmoil from Dad's death and Mother's distress. Mother and son were stuck in a vicious circle. She saw clearly that Jason was floundering and needed her more than ever, but the harder she tried, the more Jason pushed her away.

As they built a new life, they took on the challenge of using interactive media to make their family life better. Although she knew it intuitively, mom came to understand clearly that Jason's best development required full human relationships, and that his interactions with machines were no substitutes. She decided that the only role for interactive media devices in her home would be to provide opportunities to improve their collaboration and steer Jason's development.

Mom decided that it was her proper role and responsibility to change things whether Jason liked it or not. So she undertook a deliberate and thoughtful effort to join in and firmly shape Jason's interactive media world. Mother and child together creatively transformed interactive media electronic devices into assets and found new ways of appreciating each other's company.

Ever since, I have paid closer attention to how children use online computers and other interactive media. As kids use such devices, they are themselves changed, sometimes in important ways. What I discovered raised these questions:

> How can we use today's interactive media to truly improve children's development and family lives?
>
> How can we guide parents and children in their use of interactive media by applying the same basic knowledge, experience, and clinical thinking that I use daily in my work as a psychiatrist?
>
> Why not help parents realize that they can transform their online computer into a strong, rich asset for advancing their kids' development and family life?
>
> All interactive media are becoming less and less distinct, so all such devices must be included in our thinking.

I use the terms *interactive media* and *media* broadly and often interchangeably to refer to any technologically sophisticated device that produces some output in response to some behavior. Media accept (via touch screen, keyboard, game controller, dance pad, etc.), store, and process information, and in turn quickly respond by conveying sensory information that can elicit an emotional, intellectual, and/or behavioral response. Currently, these include the Internet, video games, DVDs and videos, reading devices, mobile phones, and others. Moreover, all of these interactive opportunities are being ingeniously combined into cheaper, more portable, and smaller packages such as smartphones, and new innovative devices and applications appear daily.

Music, text, or video storage devices–digital video recorders, reading tablets, MP3 players or flash drives–accept and store information and provide it back, often repetitively, any time and any place. A computer engages the user in more complex interactivity, accepting direct input via a human interface device, processing it complexly, and outputting it in a variety of visual, auditory, and mechanical ways. Moreover, an online computer provides access to vast stores of input from a great variety of people and opportunities to communicate with them in real time.

A WILD WEST

The Internet, cyberspace, baby videos, social media, cloud computing, eBooks—new words for new worlds. Our ever-expanding, vast frontier—unexplored, enthralling, dazzling, and so full of potential. A thrilling new Wild West, all ours for the taking! Technology is a fast-moving river, sweeping us along in its rapids. We are exhilarated, though we're barely steering and sometimes hardly staying afloat. We hit upon many more uses every day.

Over the past several decades, we have been amazed by the torrent of communication technologies that have burst into our lives as magical portals that take us almost anywhere with fantastic speed, enormous reach, and limitless accessibility. Innovation in access, size, portability, and human-machine interfaces are evolving with such rapidity that we can hardly imagine the interactive landscape we will experience ten years from now.

While we think of today's Internet-linked computer as the principle gateway, its functions are more and more expanded and blended with other media. A cell phone provides similar opportunities to communicate in text, video, or voice. These electronics are increasingly packaged together wirelessly, and the future will bring more innovations in their size, placement, and means of processing, making them even more pervasive and integrated into our lives

Looking back, it is easy to see how much technology has impacted us, and how fast the changes have come. The Web offers vast virtual commercial and entertainment malls, where adults and older kids naturally congregate. It is a universe we can search instantaneously. Huge virtual libraries and databases for all sorts of knowledge, reference and free press thrive there. It is a meeting place where values, politics, and spiritual and religious life simultaneously boom and clash.

We think of digital technologies as tools working for us, and indeed they are. That is why we invented them in the first place. But at the same time, technologies are doing many things to us. They are changing us profoundly, and not only by making communication, commerce, and science easier.

THE CHALLENGE

Much of this impacts our family lives and the lives of our babies and young children. And this is only the beginning. Adult computer engineers invented the first computer games for their own entertainment about sixty years ago. Since then, digital media applications have been filtering down into younger and younger age groups. Today, practically all young parents who have grown up with digital media find it natural to welcome these technologies into their homes. As a matter of fact, while "geeks" and "nerds" had been early adopters of new technologies in the past, a recent study has shown that ordinary families are now the main early adopters. And families spend less time together, especially if they have the Internet.

More than ever, media are flooding children's lives and are becoming progressively more interactive. Television has always been a somewhat interactive medium. TV engages the viewer with visual and auditory information. The viewer pays attention: learns, feels, and then responds by changing her behavior, often buying an advertiser's product. Sometimes, when excited, a viewer may yell or cheer at the screen. Some TV shows for young kids even pause the action to allow kids to react. Today, the hybrid Internet-TV marries the two powerful technologies.

Older kids invent interactions with digital technology, and industry obliges. Applications are working downward into the lives of younger and younger children. We can predict what is about to come into the crib by what older kids are doing. Currently, babies are impacted by television and DVDs in countless homes the developed world, and intelligent toys are increasingly populating their cribs.

Hardware, software, and online sites and services are becoming increasingly available to children. Digital learning applications started for older kids and adults.

Now babies are playing educational computer games. Entrepreneurial merchandisers are offering media gadgets and content and are targeting younger kids and their families.

We know that technology, as it leaps forward, will bring us things we have not yet dreamed of. But we also have a huge body of knowledge that tells us that the needs of children and families will not change. This book offers you a comprehensive system to give your children the healthiest habits for their lifetimes.

Where is all this going? Research and data are hard to come by in infancy, but there is plenty of heated debate because there is so much money involved. I started thinking about writing this book when I heard numerous good parents say things like:

- "Baby videos and TV are controversial. We need a savvy new Dr. Spock to give us guidance we can trust ..."
- "We are overwhelmed by all these devices and DVDs. Is all this good for our infants and toddlers?"
- "We like the Internet and want our second-grader to get its benefits. But everybody is talking about safety, safety, safety, and nobody is really making positive sense about the Internet working for our kids."
- "We have a lot of conflicts at home about electronics. Cell phones, MP3 players, games, flash drives, multi-function mobile phones, the Internet. Can someone who really knows please help us manage all this interactive electronic stuff for our preschooler?"
- "My eight-year-old daughter is starting to talk on the cell phone. We sometimes argue about that."
- "Our son is glued to his video game. I am afraid he is becoming an addict. How do I handle this?"

The truth is this: There is relative chaos in the digital lives of kids and families. There is no coherent and sensible, approach for parents who primarily end up reacting, while children forge ahead into uncharted landscapes. We have been thinking, "They can do what they want, as long as it is safe." But is that enough? Are we thinking carefully enough about how technology impacts our kids?

Increasing concern leads to the same question, "How can we fit new technologies into family life and raise good kids prepared them to live decent lives in a technology-rich world?" It is the right question at the right time.

Even though they have grown up with digital media, many young parents don't know how to apply them constructively in raising their children. Also, they many are uncomfortably aware that merely restricting or monitoring youngsters' media consumption is not enough. Some are beginning to realize that they must become committed to staying involved and current in order to guide their children as they develop their social and moral lives and educational and entertainment experiences.

Who would not agree that technology has made the world a different place now than it was in our childhood? But is it actually all for the better? What about the quality of our lives? How are media affecting the development of our children? What about how they experience their world, themselves, and others? What about longer term effects on values? The ways families work? Scholars are just beginning to ask these questions.

What about how we are shaping the future of society? The hand that rocks the cradle rules the world. Will it be a robotic hand? How much deliberateness, control, and thoughtfulness have we been investing into shaping how technology affects us? The answer is very little. We have mostly let ourselves be swept along and then react, just as we have to global warming, stem cell research, cloning, and nuclear energy.

Some argue that technology has not done anything to us, that we are doing it to ourselves—both the good and bad. They are right. We are in charge, and we can do better. We must prepare our children to live in a technology-saturated world and to learn to identify what is good and bad in it and how to choose the good. We can take more control for our family's sake, kid's sake, and for our society's future as technology sweeps us forward.

THE SOLUTION

Let's use these awesome technologies to enhance our family's well being and our children's development.

A healthy family is a dynamic, ever-changing living thing that adapts constantly to the social and material world around it. For many parents, good kids are their ultimate contributions to the betterment of mankind. Keeping a family healthy and strong has never been as challenging as today. And managing media is a critically important aspect of parental leadership. Here are the concerns I try to address in this book:

- How can we use today's technologies to improve our family life and children's development?
- How can we apply sound knowledge, experience, and clinical thinking to the collaborative use of interactive technologies in our homes?
- How can we harness the power of new technologies to raise better kids in better families and prepare them for life in a technology-rich world?

This is what we need: a systematic, positive approach to assigning proper use for media in our homes. We already know that establishing sound family practices and habits of mind early can lead to more successful development later on. We already know that establishing family norms and expectations improve children's lives in the challenges of adolescence. Worrying about safety has preoccupied us for over a decade. In addition to filtering and blocking methods, sophisticated ways for assuring safety are emerging. Now is the time to stop being scared and shift our thinking to a positive partnership with media.

Like many other important innovations (for example, nuclear power, industrialization, new drugs), we would be wise to anticipate that digital technologies will bring unanticipated harmful effects along with their enormous benefits. We and our kids would benefit if we spend some time carefully weighing the benefits vs. risks. And if we don't know enough, I believe it best to follow the dictum, "First, do no harm."

Here is what else you can expect in this book:

- You will take charge of your family's media, as you already do your automobile and other appliances.
- You will set proactive, positive goals.
- You will reform dramatically how your kids spend time.
- You will create a new environment around the interactive media promising mutuality, fun, and development for the entire family.
- You will redefine the role of the online computer and other gadgets in your home and adopt a brand-new, powerful framework in raising your children as they interact with streams of information from newly-available media.

A Media Plan is similar to the diet control you set for your family by selecting proper foods.

- Using the familiar analogy of what they need for good nutrition, you will learn about the Media Plans children need.
- You will determine the best mix of media and schedules appropriate for your children's healthy development, and how to use media experiences to help your children build strengths, rehabilitate deficits, and improve self-esteem, even if your child has special needs.

You will make your entire family's media experience the healthiest and most development-promoting it can be.

- You will turn media into tools that enhance the quality of life of your entire family.
- In a flexible, balanced, sensible, and positive way, for the first time, you will use media to cement your family ties and promote your children's education and overall development.
- You will have more fun with your children. From the baby on your lap to your third-grader, you can end up closer to your children. You will learn from one another and collaborate in new ways.

You will learn to stay up-to-date. You may be able to find more suitable products, services, or web sites elsewhere than cited in this book. In fact, by the time this book is in your hands, many of the specific citations here will have changed, merged, or disappeared. But my approach is specifically intended to serve you even as technology jumps forward. The more work online and exploring you do, the better.

- Since web sites come and go, I provide you with the site **www.mydigitalfamily.org** to supplement, update, and provide easy means to help you.
- We also offer you the means of searching for the latest sites yourself using a string of key search words. You may create your own string of keywords from text or headings in this book and use search engines to find suitable sites. For example, *children development computer furniture learning* will yield fascinating online discoveries. Try it.

Compared to many available guides devoted to safety, this book focuses intensely on the family.

- You will achieve the right balance of innovation and caution as you move ahead with our children thoughtfully and carefully into exciting frontiers.
- You will learn useful and durable ideas to help you keep your family safe.
- You will learn how to prevent and cope with a child's Internet addiction and how to protect your family's well-being.

Bottom line: Expect to become a better parent. You are the only parent(s) your kid(s) have. You are planting seeds that will eventually germinate and blossom into powerful influences later in your child's life. Your commitment now will pay dividends later.

PART 1: SETUP AND QUICK START GUIDE

Congratulations! You are about to welcome—or have already welcomed—your brand new baby into your home. She needs you to enjoy her. As time goes on and she makes her way to becoming the person she is meant to be, I am sure that you will appreciate even more this wonderful event.

Baby is a brand new family member right "out of the box." This book is your instruction manual to understand and manage your young children's media life, and in the process to deal with much else in their development and your family life. This first section, *Setup and Quick Start Guide,* will help you set up and perform routine media-related tasks right out of the box. The *User's Guide* that follows will provide you and your professional advisors with more extensive tools.

Your baby is a complicated and delicate creature, yet quite resilient and ready to do her only jobs—thrive and develop. Nature has already largely prepared you for your part, so trust your intuition, at least as a starting point. Your own natural skills can provide you with much of the guidance you will need to make baby's stay in your home a blessing and a success. You can also rely on other parents and family members—especially experts like pediatricians—and baby's own reactions.

But you will often wish you had careful advice, reassurance, and instruction manuals, especially in areas new to our civilization—like interactive media. As Baby enters and grows up in an increasingly technology-intense and complex world that is changing faster than most of us can keep up with, this book is your instruction manual for baby's psychological and emotional development in this evolving environment. Here you will learn how to best understand and position baby in this important part of her world so that you can safeguard and enhance the delicate environments of her brain and mind.

Chapter 1:	**An Extreme Makeover—Appreciate That You Are Wiring Your Children's Brains**
Chapter 2:	**The Media Plan—Harvesting Growth Opportunities from Media**
Chapter 3:	**Setup and Installation of the Media Plan**
Chapter 4:	**Keeping Your Baby's Brain "Green" (Birth to Two)**
Chapter 5:	**Learning What's Real (Two to Five)**
Chapter 6:	**Collaborating with Your Kid (Five to Eight)**

Chapter 1: An Extreme Makeover—Appreciate That You Are Wiring Your Children's Brains

We all recognize the allure and power of media. It is likely that without much planning on your part, media have already led to changes in your home life and in some of the ways you interact with your children. In general, the earlier you start making the healthy use of media a normal part of family life, the more you will be doing to improve it now and in the future.

My emphasis is on assisting families to raise healthy and productive kids. As you get your baby "out of the box," this book is your new instruction manual for operating media in your home in the best and safest way. By learning to take full advantage of the benefits of this approach, you will inevitably also learn to increase your general effectiveness as a parent.

SOME RECENT STUDIES

Studies are beginning to appear in respected scientific journals. They suggest that media consumption among children is pervasive, and that early childhood patterns persist into later years. One team concludes, "These children are growing up in a media-saturated environment with almost universal access to television, and a striking number have a television in their bedroom. Media and technology are here to stay and are virtually guaranteed to play an ever-increasing role in daily life, even among the very young. Additional research on their developmental impact is crucial to public health."

Another team reports that the more time kids spent watching TV, the less time they spend with family members, doing homework, and playing creatively. Still another team from San Diego State University recently reported that high users of media at young ages are likely to remain high users when older. Adolescent gamers playing video games without parents or friends spent less time with family or friends in other activities and about a third less time reading and doing homework.

These findings, if confirmed, are alarming. Uncontrolled media consumption could destroy healthy development of children and healthy families. Coupled with recent reports of our failing educational system, do you want to take these risks with your kids?

GOOD FAMILIES MAKE BETTER BRAINS

Believe it or not, you are actually gradually programming circuits in your child's brain every day. Let's start at the beginning and review what we know.

Birth—what a miracle! All parents have been awed by this most amazing process. Your precious newborn bundle of joy lying on her back, occasionally flailing her arms, looking, burping, sleeping, smiling … full of promise and potential, and utterly dependent on you. You are fully focused on the moment and don't yet grasp all that is coming.

What's coming is more of the same miracle. Your baby will spend her next twenty years gradually learning, and separating from you, and individuating into the complex, integrated, social, intellectual, moral, physical, and aesthetic human being she is meant to become; able to make her own destiny with freedom and responsibility. Before you know it, she will blossom into a fully-grown person with her own life, family, career, and children. Human development is a miraculous orchestration and much more than merely the sum of its ingredients.

Of all creatures, mammals are the most developed socially, and we humans are the most developed of all mammals. What makes that possible is our human brain. Some scientists now believe that our brain is probably the most sophisticated object in the universe, as it keeps evolving primarily to enable us to live and survive with each other in social groups. I personally firmly believe that brains did not evolve to do rocket science, but mostly to enable our survival with each other. Survival of isolated creatures in the wild does not require the complexity of thought and feeling that living with people in social groups does.

For our brains to fully mature and learn social- and self-awareness, understanding of others, and skilled behavior, we humans require a long maturational and developmental period. Even with recent lengthening of life spans, our basic developmental period can still take up to a full third of our lives.

To complete the picture, we must realize that the process of development is never really over. In addition to infancy and childhood, adolescence is also a time for a major developmental surge. Like those of many other mammalian species, human adolescents show more risk-taking, impulsivity, as well as novelty and sensation-seeking; increased interactions with peers and focus on peer-directed socialization; gender-driven behavior; and conflict with adults.

During this developmental surge in preparation for adulthood, your own adolescent will explore the limits of conventional expectations and assumptions and the limits of his abilities. Underlying all this are physical changes as his brain becomes more finely tuned and efficient under the influence of genetic and hormonal changes, allowing for more effective processing of information and regulation of emotion. However, scientists have recently discovered that throughout life, brain maturation can continue to be influenced by the environment. Other features of adolescence include uneven development and erratic and inconsistent behavior. When mixed with experimentation and the blossoming of sexuality and gender identity, high-risk behaviors can exacerbate family conflict. Whatever the circumstances, however, good family life from infancy provides the best brain infrastructure for later development.

Nature gifts every baby with a personal master plan and timetable to become who she can be physically and psychologically. Each of her body's organs is coded with its own blueprint and schedule for growth and maturation in a marvelous and complex choreography. Baby's brain is the central organizer for this magic. This organ's blueprint calls for interaction with the environment to enable it to select a particular design. Baby needs to touch, be touched, hear, see, move, taste, smell, feel, eat, be loved and responded to so that she can learn how to process and use these modalities and develop ways of making sense of them and herself. Nature is clever—she does not abandon Baby to random inputs. She gives each baby her own plan and timetable to select the best stimulation that will enable her brain to achieve its best growth. All of this is prewired and prepackaged.

Baby needs her parents' help to accomplish her tasks. Baby relies on her inborn design to choose what she needs, when she needs it. She counts on her parents to do the basics—feel hope, responsibility, love, and give protection and nourishment. Baby also counts on her parents to carry, feed, clean, hold, rock, talk to, cheer, be fully present with, and play with her. In an average environment, Baby has plenty to choose from, as she pursues her natural tendency to mature a brain that has a balanced ecology. However, we don't really know much yet about the specifics of what she needs and how exactly Baby senses, selects input, learns, and maintains the overall balance of her brain's ecology.

Nevertheless, it is our task as parents to preserve this ecology of Baby's mind, and certainly not to distort or overload it. It is our job to always think of how to preserve its balance with the same care and sensitivity we give to thinking about preserving the world's ecological balance.

First, do no harm. This principle must apply to how we place interactive media in Baby's world. In addition to protecting Baby from harm, I recommend that we try very cautiously to improve her world with interactive media. I believe we can best guide ourselves by evaluating media in terms of the Growth Opportunities offered our babies and young children.

PARENTING BASICS

NEWSFLASH: READ ALL ABOUT IT

Breaking News: Family Hardwires and Programs Twenty-Year-Old's Brain

Parents are actually hardwiring new circuits in kids' brains every day. In an international conference in Chicago, scientists are now showing that interpersonal environments actually often determine the physical brain's workings throughout life, but especially until young adulthood—the early twenties—by modifying how genes operate in the brain. They confirm why we need good families: Families alter the hardwiring and software of the child's brain through adolescence.

Using the latest technology, neuroscientists are beginning to confirm that actual physical events have underlain all along earlier findings and concepts of renowned pioneers like Mary Ainsworth, John Bowlby, Eric Erickson, Anna Freud, Sigmund Freud, Harry Harlow, Heinz Kohut, Margaret Mahler, Jean Piaget, Michael Rutter, Rene Spitz, and other students of the developing mind and behavior.

Scientists explain: Imagine that your baby inherits from you a huge library, made up of many rooms, each containing shelves and shelves of blueprints she will use to construct everything about herself. Each parent contributed half of the library. Each blueprint is a plan for a small part of how and when she will be built physically, intellectually, and emotionally. Some rooms and plans are more easily accessible—like plans for healthy lungs, heart, and skin.

To get the blueprints for building other parts of her, like her emotional makeup, intellectual capacity, and special talents, she must get the blueprints that are locked and guarded in other rooms. She can get a passkey to these special rooms or wings of the library only if she has had certain experiences. For example, being loved and protected as an infant is especially important for her to get into the wing where the blueprints for a healthy and productive person are housed. If she had been abused, she will be forced to use blueprints that could lead to emotional turmoil in her life. If she is not sufficiently loved and protected, she cannot get the passkey, and must do with more basic and sometimes inadequate plans for her growth and development.

DNA is the library that contains the inherited blueprint genes, and environmental experiences can allow access to various parts of the molecule. According to one neuroscientist, parents can "make better brains that allow for better minds and make for better persons, parents, and citizens. Our young brains are very plastic. How the blueprints that genes provide are translated into actual functioning brain tissues is wonderfully and sensitively tuned to the environment of the young person."

A prominent neuropsychologist said, "Who could imagine a more elegant and miraculous system? These new discoveries are showing clearly how nurture actually works in the famous nature vs. nurture story. It is not a 'vs.' equation. It is a nature plus nurture story. The brain continues to mature through young adulthood. At least until the early twenties, families must buttress and scaffold the developing person in many ways to shape circuits in the brain of the youngster that enable resiliency to the effects of stress, and successful functioning within the community."

During later childhood, and throughout life, families buttress and scaffold the developing person and change her brain in many ways. Many scientists believe that the complex human brain has evolved over many millennia primarily to learn and enable social relationships with other people. Our unique abilities to direct attention at will and our eagerness to learn from our elders and each other originate in our uniquely human brains.

Upon hearing this news, an eminent religious leader declared, "God works in mysterious and wonderful ways! He gives us the power of choice between good and evil, life and death. Here's another choice he gives parents—to choose life and do good."

An eminent child psychiatrist said, "It is both mind over matter and matter over mind. Sound interpersonal interactions among family members are essential."

Another notable scientist and humanist added, "Newborns are gifts to us, prewired to promise the most intense and rewarding relationships of our lives. They engage us readily and lovingly with the best they have got. Infants blessed with sensitive, available, interactive, consistent, and nurturing mothers, persons who are themselves protected from stress by other family members and society during childhood and adolescence develop better brains that support a lifetime of interpersonal relationships, psychological and physical health, and intellectual functioning."

This is good news for parents. Without a doubt, good relationships among family members in the home are the most significant influences to assure a child's healthy development. There is now no doubt that the emotional and physical environment parents provide their children, even through their twenties, sets the physical basis of later brain circuits and networks as well as hormonal activities. We emerge best into adulthood in the safety of a home, under the protection, support, and guidance of our families. We must do all we can to support families if we want good people in a good world.

The overwhelming majority of kids are born prewired to be hopeful and optimistic, eager to give and receive love. Good families—and there are many—cultivate these precious gifts and are blessings to us all. We now have brilliant and wonderful technologies. Let's use them to make family life better.

MAKE A SERIOUS COMMITMENT

Becoming parents—and we are constantly becoming and re-becoming parents as our children grow—is actually a personal developmental journey for each of us. Studies even show that a new mother's brain itself is changed in her new role. If we reflect back upon our pre-parent years, we can now appreciate how much we ourselves have evolved as persons. Going in, we knew that our mission as parents would be to work hard at loving our babies by providing them the best and safest emotional, psychological, and physical opportunities.

Hopefully, by now, we have learned to anchor ourselves in our families and communities to enable us to cope with our demanding and complex world. We already accept that we must face daily countless decisions for which we often feel unprepared, yet have enough confidence to fly by the seat of our pants.

Many of the complicated decisions we are now increasingly facing will be related to the ever-present technological gadgets in our kids' lives. We know that technology is here to stay, and we are pretty sure it can be a tremendous blessing. But we also hardly know how it works in the lives of children and how it can help us raise good children.

You must first get it very clear in your mind that you can and will make intelligent decisions and find ways to raise good kids in our technology-rich world. Your first decision is to commit yourself to a new challenge: From now on, you will include managing your family's media interactions within your overall idea of yourself as your family's leader. You will now be aiming at assuring that your children's interactions with media become assets to their growth and development. Making a Media Plan is like making retirement plans and college savings plans – laying the foundation for a positive future.

How do we tap this tremendous potential while protecting our children from the hazards of interacting with media? The answer may seem simple, but a paradigm shift is essential. I suggest that you think of media gadgets as home appliances that can be used and misused, and can be safe or hazardous, like other appliances.

Start thinking of digital gadgets in your home as pieces of a larger system—your family's media environment. You choose energy-efficient appliances to preserve the physical environment. How about thinking the same about how media preserves your family's emotional health and the delicate environments within your child's developing brain. Think of your desktop as a family computer, to emphasize its positive role in your family life. Similarly is it a baby DVD or a family DVD? Your third-grader's cell phone or the family's cell phone? These are all parts of your family's media environment.

We use a kitchen stove to prepare anything from snacks to gourmet meals. Like the stove in our kitchen or the car in our garage, the magic boxes we call interactive media are nothing more than handy appliances -- our tools. They are merely our means to an end. And their only use that makes sense is as resources for raising our children and making our family better. There is no magic to that.

If you really buy into this concept, then you will use digital gadgets with the same thoughtfulness and care, as you do your family car or kitchen oven. So start now to include these in your daily life as parenting tools and take full control.

A good time to initiate this new way of living with digital media may be when you buy a new device, software, or a subscription to a new service provider.

MAKE MEDIA YOUR FAMILY APPLIANCES

- Realize that media and online experiences are shaping what kind of a person your child is becoming from infancy onward.

- Think of your child's interactive media experiences solely as opportunities to nurture balanced social, moral, intellectual, and aesthetic development.

- Treat your child's interactive media experiences with the same care you organize and plan school lunches, family trips, healthcare, and other family activities.

- Take charge of your family's media appliances, as you do the family car or kitchen stove.

- Make interactive media your own powerful collaborative tools, using them to enhance your children's development and your family life.

- Know your own child and what he needs as he develops. Do not rely solely on expert guidelines or testimonials.

- Pay close attention to credible rating systems for baby videos, movies, TV, Internet sites, and video games. Seek reviews by unbiased, respected experts for appropriate age and content guidelines.

- Prevent exposure to media violence, including excessive news coverage of violence, TV shows, and video games.

- Do your due diligence. Know computers, the Web and other media. Find good sites.

- Decide the mix you want for your child's online and offline games and entertainment; communication with friends; moral and spiritual ideas; geography and social awareness, and knowledge of government and current events.

- Decide if the total weekly time your child spends with interactive media in relation to other possible activities is reasonable.

- Rearrange the physical placement of the computer and its peripherals.

- Know your child's vulnerabilities, minimize hazards, and teach safety, including inoculating and shielding from unwanted media influences.

- Give your older child routine media access via mobile devices only for compelling reasons.

- Join other parents to advocate for child- and family-oriented media portals and development of supporting software and hardware.

- Support research and public efforts that enhance beneficial and responsible uses of interactive media.

- For current information, visit www.mydigitalfamily.org.

JUMP RIGHT IN

Search words: furniture home computer children family

Here we come to another major shift in our thinking—how the spatial arrangement of the computer within a room impacts on interactions among family members. If we want to change how we use the online home computer, we must change where it is placed. If you change the positioning of the people interacting with media, you quickly expand the role of the computer in your home to create a family computer that provides new opportunities for learning and connecting.

Computers are usually set up for one person alone, facing away from other people, usually against a wall, with accessories for only one person—one keyboard, one display, and one mouse. As a workstation or for individual concentrated work, this typical linear configuration may be ideal. But the mindset it creates is narrow and solitary. The array enables interaction between the user and the machine, but it shuts out other direct human interaction. In that sense, it is basically an antisocial, or at best, asocial, physical arrangement.

The current primacy of man-machine dyad should give way to human-to-human interactions whenever possible in our homes. No wonder the use of computers is a source of conflict and power struggles in so many homes—we are confusing our kids.

Children tend to think literally and see the world we arrange around them as loud and clear non-verbal statements. When we close the bathroom door, we are teaching privacy. When we put the home computer against a wall or in the child's room, our message is "We arranged the placement of our computer for a reason: Your computer time in our home is private and solitary. It is your business and part of your private time. It is not a family activity and has no value as such. It is OK with us, your parents, that you shut us out and immerse yourself in your online activities. We are saying, beyond maybe setting some safety limits on what you can do online, we really have no other role in your interactive media experience."

Look at the computer in your own home. Does it enable or block interpersonal interactions? Fix it so it works for more than one solitary person, even before your baby arrives. Then, sit with Baby and very young children on your lap. Turn the child so that you can keep good eye contact. Parents can interact with each other and Baby through software and specially selected Internet sites as well as through other input and output devices. The furniture arrangement should enable two or more family members fully interacting with each other.

Later, change the arrangement to a triangle. The child can be in a high chair and you would face each other, assuring maximum access and face-to-face contact. By pulling the device slightly away from the wall, adding a larger monitor, and putting several chairs in front of the computer, you will now enhance interpersonal interactions in your home. Many systems allow you to install a wireless keyboard with a trackball mouse and work parallel to your child. If you use a laptop, connect it to a larger display so that more than one person can see the screen clearly.

When deciding how to place the online family computer in your home, also consider that much of your older child's online time will require your supervision. You want to place it close enough to converse with your child and see the screen for yourself. Place the computer for your convenience, so that you can pursue your own activities near the child whenever he is online.

Similarly, get your kids used to listening to music together with other family members via speakers, rather than isolating earphones. The key is being present with each other.

PARENTING BASICS

BEING FULLY PRESENT TOGETHER IS THE GLUE

Being fully present together with the child means being actively in the here and now, connected, plugged, and tuned into the moment of being with the child. Being fully present with your baby is speaking, listening, smelling, touching, and looking together. Parent-child interactions commonly involve doing for, doing to, and doing with each other. But as far as healthy emotional development goes, being fully present together is the most important component of these interactions. Social bonds and moral or ethical behaviors remain underdeveloped if parents are not fully present with their children.

Relationships are what make us human - the family is the launching pad for all relationships. Being fully present together is so important that our babies are prewired to engage us with their reflexive smile from their first moments, and we are prewired to respond to them (witness how people react to a baby on an elevator).

Better kids are made by parents who are fully present with their children. It takes a special and ongoing effort, motivation, and commitment. It does not necessarily always come naturally. It is demanding for busy parents distracted with everyday pressures but is a vital nutrient for raising good and full persons.

The younger the child, the more important eye-to-eye contact is, along with touching, holding, and talking or singing. Eat together and be present with each other at least once a day, sitting down face-to-face around the table, free of all distractions. Turn your baby's stroller seat so that he faces you, and keep good eye and voice contact with him; keep your older child free of excessive out of home activities; use car rides as occasions for being present together—even in silence, certainly in conversation or singing, rather than letting the child interact with media in the backseat. Bend down to eye level when talking to your child, or pick up the younger child to make face-to-face contact as often as you can. I am now inclined to say to parents: Be fully present with your children. Ban distracting media from your family time in the car and at home. When distracted online, a phone call, or texting, you may be interrupting the vital bond your kids need for healthy wiring of their brains.

Pay attention and be aware of the countless opportunities to be present together in everyday life. Your child will learn from you. A word of caution: Being present does not mean over-involvement, intrusion, and hovering. It means being together, each aware of the other as a person in her own space. And by the way, children are usually unaware of what is missing. So, don't count on them to alert you to their needs for your full presence.

Something's wrong? In my practice and elsewhere, I see more children today—some very bright and from good homes where they have been kept busy with all sorts of worthwhile activities—who are underdeveloped as people. They are disconnected, flat, listless, easily bored, inarticulate, and often passive. These children seem emptier, more distant, and their gaze more vacant. They cannot tolerate being by themselves or without a certain level of often mindless stimulation. They glide through superficial relationships as if coated with Teflon and have weak commitments and values. They often lack ambition, direction, motivation, and a twinkle in their eyes. They often complain of boredom. These children are less capable of being fully present within themselves and in their own lives. They miss out on the pleasure of collaboration with others and fall behind in other social skills. They do not know how to be present in their own lives.

I have wondered if this is in some way caused by a trend that includes the increasing solitary exposure to interactive media in their homes. Parents and kids can more easily become distracted away from each other and drift into becoming less present together. These children and teens remind me somewhat of children raised in poorly run orphanages, kids who had very depressed, anxious, or otherwise unavailable mothers or principle caretakers, kids from very troubled homes, and of foster children who had too many placements. These children form superficial relationships and seem to lack enough glue to hold their personhood together. Their selves are enfeebled and incomplete. They are drifting and lost, long before they come to our attention as underachievers, suicides, or school shooters.

It should be pretty obvious by now that with a little planning and some rearranging you can make the computer and other digital devices wonderful assets in your home. You will see in later chapters that exciting possibilities are endless, limited only by your motivation and imagination. By using digital media in this way, you will be teaching your child that we do have control over technology and that it can be a good thing in their lives if used well. Children will surely take this critical lesson into their technology-rich futures.

Chapter 2: The Media Plan—Harvesting Growth Opportunities from Media

We have an obesity epidemic in the United States because too many children are left to themselves to consume unhealthy market-driven diets. Similarly, the consequences of leaving our children to themselves to consume media are not good for them either.

There is no doubt that interactive media will continue to have a growing place in the lives of families and in the development of children. The type and content of media available to babies and young children proliferate at astounding rates. However, there has been little in the way of a coherent system to understand and manage this phenomenon. In this chapter, you will learn a framework to think about this critical area of family life.

In this section you will learn how to determine the best mix of interactive media nutrients for your children, much as you create healthy menus of proteins, fats, carbohydrates, and vitamins and minerals every day.

Growth Opportunities

The "Food Groups" of the Media Plan

- Family Relationships
- Socialization
- Values Education
- Education Enrichment
- Entertainment

<u>*Basic Principle*</u>: **Interactive media must always promote healthy family life and child development. Otherwise, it makes no sense to have it in your home.**

THE MAIN PURPOSE OF MEDIA IS TO PROMOTE HEALTHY DEVELOPMENT AND FAMILY LIFE

Each developmental phase builds upon the one before. Scientists and child advocates have established the basic needs of older children. As our kids evolve from newborns through infancy, childhood, and adolescence, what happens in our homes defines them as individuals and, ultimately, will determine the nature of society and the course of civilization.

We have general expectations of what is normal as the child moves forward along her developmental trajectory. But the process is unique for each child and family. The trajectory is never smooth, but rather a unique combination of small oscillations and large plateaus interspersed with rapid changes, and occasional reversals, reflecting periods of acceleration and consolidation.

All aspects of development are closely interrelated, and the whole is much greater than the sum of the parts, in the same way that great music is much more than a mere sum of its notes. I have never ceased to be awestruck by the elegance and grace of how brain events and developmental tasks blend together in a wonderfully choreographed dynamic process that makes a person.

My advice: Enjoy your children and be grateful to them for who they are now and for giving you a chance to participate in the miracle of human growth and development.

Kids, especially babies, are not little adults. Their needs and the realities and worlds they live in day-to-day are vastly different from those of grown-ups, and often only dimly understood by them. A good childhood is not merely preparation for adulthood, nor is adolescence just a bridge between childhood and adulthood. These phases of life are in themselves rich and challenging ways of being human.

Take a moment and remember. Reach back to your own childhood or teenage years and remember how you felt then about yourself and the people and the world around you. Look at photos of yourself as a youngster. You will realize that preparation for adulthood was the furthest thing from your mind.

Scientific information is coming slowly, and research on the effects of interactive media is in its infancy. Many of the findings are still contradictory and confusing, but initial studies suggest that interactive media are a mixed blessing for the development of children.

PARENTING BASICS
WHAT DOES "HEALTHY CHILD DEVELOPMENT" ACTUALLY MEAN?

Sensorimotor and cognitive development, also known as neurocognitive development, refers to the physical maturation of the brain and concomitant development of its functions and processing abilities. Its aim is to help children achieve their physical and intellectual best. Some of these brain functions are: ability to learn language, pay attention, comprehend, manipulate symbols, process complex information, learn, remember, and coordinate expression in movement or language in a smooth and integrated manner. The main purpose of education is to maximize this aspect of development.

Emotional development evolves from a stable, mature, and independent personality. Feelings flow in increasing complexity and richness, and are regulated and harnessed to set the tone and energy for other functions.

Social development makes a person capable of living productively in a community, finding intimacy, and practicing social responsibility and even leadership. Socially, a young child slowly moves from parallel play to interaction-rich play, from concrete play to abstract thinking and creativity. Collaborative thinking, empathy, sympathy, learning, and teaching are social activities essential to human social and intellectual life.

Moral development aims at developing a moral compass, achieving a refined sense of right and wrong, and living accordingly, often guided by religious and spiritual systems.

Self-care skills development allows for better caring of the body, moving toward a lifetime of good health practices, including the value of regular exercise. We teach our children basic hygiene from the beginning when they have their first teeth and we explain and demonstrate how to brush teeth. These skills begin in the home in early childhood and progress through youth and even well into adulthood.

Aesthetic development targets the richness and beauty that life offers through the senses and physical movement and their influence on the mind.

Self-development results in a person who can be together with others. This includes achieving and maintaining a good balance among ideals, ambitions, and intellectual, social, spiritual and emotional functioning. The whole self is an almost seamless integration of these components and is greater than the sum of its parts.

MEDIA PLANS

One aspect of loving children is providing healthy nutrition. We give our children proper and balanced foods to assure their physical growth. We also carefully nourish them with opportunities to socialize, play, and learn, maximizing their personality development. Merely restricting the foods our kids eat is not enough, and neither is merely limiting their social or academic opportunities.

Besides coming from different places in nature, food groups differ in the proportions of nutrients they contain. A diet is a special nutritional plan based on bodily needs designed to guide us in providing the basic nutrients that enable our physical health, growth, and maturation.

These basic nutrients must be in the correct amounts and balance—too much or too little disturbs the body's equilibrium. The basic nutrients of foods include proteins that promote growth of muscle tissue and other necessary bodily processes, fats that insulate the body and store energy, carbohydrates that fuel the body and give it energy, and vitamins (A, B, C, D, etc.) and minerals (calcium, potassium, sodium, etc.), also necessary for overall healthy functioning.

In the same way, loving your kids must include feeding them a balanced Media Plan, and not merely restricting media consumption. Think of the right mix of interactive media as a healthy diet, relying on what we know as good for children in enhancing intellectual, social, and emotional development.

When we create a food nutrition plan for our family, we first think of what they need and then of what we can get in the grocery store—foods that belong to food groups arranged in the well-known pyramid to make a proper diet. In addition to maintaining health, we look to nutrition plans to promote physical and mental maturation.

Similarly, we look to the Media Plan to enable emotional and psychological development. Just as children learn to consume food in the context of a family, so with consuming media. The Media Plan you will create is built around a menu of opportunities present in media to enhance development, very similar to the mix of food groups of a healthy diet.

Growth Opportunities are the building blocks of the Media Plan. They are the food groups, and include Family Relationships, Socialization, Values Education, Education Enrichment, and Entertainment. You harvest Growth Opportunities from interactive media like the Internet, just as you shop for your family's meals according to the food groups. It makes sense to include planning packed lunches, picnic baskets, or restaurant meals bought outside the home as we design diets. Similarly, stand-alone or portable interactive media devices like smartphones that provide interactive opportunities are all part of the same system. You will even include television viewing in your planning.

Here are some principles:
- All media content must include at least one Growth Opportunity.
- Appropriate type, amount, and quality of content should match the child's age and help him stretch his abilities to progress along each developmental trajectory.
- The form of interactive media should match a child's ability to perceive, pay attention, and learn.

THE GUIDELINES

How much, how often, what, and when we consume are important aspects of good nutrition plans. In this book, the unit of media consumption is time per week, excluding school work. I give Media Plan guidelines according to the child's age. The total time is divided among Growth Opportunities in several daily sessions. An important aspect of the Plan is interaction with family members and the presence of parents.

Guidelines in this book are merely thoughtful suggestions, often based on my own observations, readings, and clinical judgment, usually without actual rigorous science to back them up. That is the nature of clinical decision-making. For example, in prescribing a medication, physicians start with a medication's general guidelines and then adjust the dose to fit the patient's unique circumstances. Same here—these are the general starting points.

For example, I recommend that two-year-olds spend no more than one hour weekly with interactive media, such as computer games, with another two hours or so for passive activities such as other screen time and listening to music or radio (together with a parent at all times). At the other end of our age range in this book, eight-year-olds can have a maximum of five hours a week of non-school interactive media time and about another five hours for other media.

Part of healthy eating is spacing consumption throughout a day. Similarly, the Media Plan works best when consumption is divided among several sessions during a week. I therefore recommend that parents limit time spent in any one session, especially for entertainment, gaming or using social media. Starting with a maximum of fifteen minutes for young pre-schoolers, I suggest that an hour per session is plenty for eight-year- olds.

Parents who wish to keep media out of their kids' lives altogether might nevertheless consider a minimum of one hour a week in their presence by age eight to give the child an opportunity to learn that media can be a healthy aspect of family life.

FAMILY RELATIONSHIPS

Interactive media offer opportunities for engaging two or more family members in a satisfying mutual interaction that strengthens their relationship and their appreciation of their importance to each other. Sound, interpersonal interactions among family members are essential. Without a doubt, good relationships among family members in the home are the most significant influences to assure your child's healthy development. The younger the child, the more all other Growth Opportunities are directly tied to Family Relationships.

Family Relationships is the most essential ingredient in Media Plans throughout youth. This crucial component is expressed directly through both the content of activity, as well as the interactions among participants. In the early pre-school years it is the only Growth Opportunity of any consequence offered by interactive media.

Parental participation in interactive media activities creates the psychological and emotional foundation for the entire interactive media experience. An infant does not belong in front of a screen without a parent's presence. The younger the child, the more present you must be, participating in activities appropriate to the child's needs. Interactive media provide easy and fun opportunities for connecting with each other and learning to plan and think together no matter what the child's age.

The spatial arrangement of input and output devices should maximize opportunities for positive interactions among people and provide optimal challenges to motor and sensory development. Several family members can enjoy the online and interactive media experience together as they talk and look at each other. They interact via the machine, much as they do while playing a sport or game or watching a movie together. Such joint activities stimulate novel and varied experiences that also underlie good cognitive development. The key is family interaction.

Multiple human interface devices (HIDs)—keyboard, mouse, joystick, dance pad, motion detector, microphone, Webcam, or touch screen—can promote simultaneous interactions. Output devices are usually display screens and speakers or headphones. In the future, HIDs may stimulate other senses, like touch, temperature, position, pressure, or even smell, especially for people with special needs.

Interactions among family members around interactive media can be enriching for all: For example, Mom learning to make word sounds with Baby using a baby video as her guide; Dad figuring out how to use a new device or game together with a four-year-old; looking at pictures with a 2 ½-year-old; a look at Google Earth with a third-grader; or learning arts and crafts projects together.

If you want your family's meals to affirm family life, do your best to prepare have mealtimes together at least once daily. Meals offer opportunities for social gatherings and all the interpersonal richness that comes with them. A comfortable and attractive setting affirms the value of the family. Often, a meal prepared with careful attention to the details provides special opportunities for valuing the companionship of family members. Involving children in preparing the meal and setting the table adds to their sense of belonging and responsibility. Additionally, family-style meals have a different feel than those individually served—more communal, warm, and intimate.

NOURISHING YOUR KIDS: A SIDE-BY-SIDE COMPARISON

A BALANCED FOOD DIET PROMOTES CHILDREN'S PHYSICAL GROWTH AND GENERAL HEALTH of brain, nerves, muscles, bones, cardiovascular, blood, glands, etc. by balancing ESSENTIAL NUTRIENTS fats, carbohydrates, protein, vitamins, minerals	A BALANCED MEDIA PLAN PROMOTES CHILDREN'S PSYCHOLOGICAL HEALTH AND GOOD DEVELOPMENT of neurocognitive, sensorimotor, emotional, social, moral, self care, aesthetic abilities by balancing ESSENTIAL INTERACTIVE MEDIA NUTRIENTS parent participation, appropriate content, learning how to learn, pleasure, play
MEALS ARE ORGANIZED BY FOOD GROUPS meat, dairy, fruits and vegetables, grains Grocery items often contain several food groups—a beef stew has both meat and vegetables.	INTERACTIVE MEDIA ARE ORGANIZED BY GROWTH OPPORTUNITIES Growth Opportunities: Family Relationships, Socialization, Values Education, Education Enrichment, and Entertainment Interactive media often offer several Growth Opportunities
MENU PLANNING, SHOPPING, FOOD PREPARATION, AND MEAL PRACTICES Express the attitudes, values, and general health practices of the family, community, and culture and unique needs of the child. Get all family members to sign on, make it fun and get your child to help.	MENU PLANNING, INTERACTIVE MEDIA SELECTION, AND HOME PRACTICES Express the attitudes, values and psychological health practices of the family, community, and culture and unique needs of the child. Get all family members to sign on, make it fun and get your child to help.

Agreeing on the basics of your Media Plans for the kids, like planning and preparing menus and gathering to enjoy the benefits together family-style, can be an enriching experience for all. Empower your older kids by including them as partners in the planning phase. Planning your Media Plan together is an opportunity for closeness. The very act of such collaborative thinking can bring parent and child together with mutual respect and empowerment and reduce power conflicts. This is good training for replacing later skirmishes and conflicts with collaboration. You might reduce difficulty later with adolescents if they grow up with this system.

As with daily meals, it is imperative to have regular family interactive media or online sessions. Innovative human interface devices or musical technologies like a fine Internet-connected grand piano and composing keyboard provide great opportunities for hours of family fun and education.

New digital products that impact family life are introduced constantly. There are many ways that interactive media can promote your relationship with your school-age child and enhance family life for all family members. You must make decisions about allowing interactive media into your home thoughtfully and in the larger context of your children's overall development.

For example, would a smartphone with specially written 'apps' for a six-year-old year-old be a good idea? There may be advantages like convenience and pleasing the child. However, what about the distractions and limitations? What about evidence that overuse of digital media can harm development and family life?

I suggest that interactive media exposure, such as computer games, might begin with one hour per week for two-year-olds, with another two hours for passive activities like other screen time and listening to music or radio. At the other end of this age range, I suggest for eight-year-olds a maximum of five hours a week of interactive media time and about another five for other media.

Being present with the child is a crucial element of effective parenting. Two-year-olds should always be interacting with media together with their parents or other caregivers. By age eight, I suggest that parents continue to spend a minimum of almost three hours a week actively interacting with children around media and another two hours attentive and nearby, leaving the child totally alone for less than an hour.

SOCIALIZATION

For Baby, Socialization begins and ends in the family. She learns how to love and be loved in the most basic ways that pull together her mind and body. With time, she expands her sphere and will spend her lifetime learning about all the nuances of love from intimacy, love for her own kids, love of her neighbor, and love of mankind.

Healthy and appropriate interactions with people at home and through media assure appropriate quality and quantity of contact with diverse people—friends, family, merchants, and others. Such contact can advance humor, tact, kindness, gratitude, compassion, fairness, leadership and consumer skills.

Although interactive media may be wonderful recent tools, basic standards of politeness are not. It is never too early to emphasize the need for civility and model polite behavior. Later, safe and honest commercial transactions, writing appropriate e-mails and text messages, or participating in online auctions are examples of the variety of ways to interact. Media must provide Growth Opportunities for interacting skillfully, appropriately, and positively with peers and adults, communicating effectively, and treating others with kindness.

Preschool children's playing together, developing skills in other social behaviors, and forming pro-social attitudes through interactive media are great opportunities for development. Make it a point to teach basic respect, waiting for someone to finish while not making others wait, and not interrupting.

Interactive media can offer wonderful venues for Socialization. Baby videos and children's TV can model pro-social behaviors. Virtual communities that draw members from the world over can cement ties that promote tolerance and peace among third-graders and their families.

Real human heroes—individuals from the present or past whose lives stand for something good—can inspire children as early as preschool. Socialization opportunities can be found on game sites, chat groups, and instant messaging. Music and video sites enrich the child's life and often provide bonds for social interactions with peers. These can all teach values, social skills, and can be fun. Online communities can spontaneously mushroom as parents of young children respond to a traumatic event. Social networks are becoming more and more community-oriented. However, terrorist organizations and other antisocial groups recruit members and spread hate, so parents must be vigilant.

Quantity is often the enemy of quality. Although the brain wiring of most normal kids in healthy families eventually drives them to seek quality, they require protection and guidance in limiting overload and noise and in selecting the right opportunities. Infants require protection from inappropriate stimulation. Social networks can begin to provide virtual play groups in a pinch, but only with parental presence. Older kids need to be taught to distinguish between the richness that comes from enduring relationships from the noisy chatter that surrounds them.

Media usage reflects gender differences. Girls tend to spend more time text-messaging friends. They also tend to continue familiar friendships rather than start new ones. Digital communication forms another link in the chain of interactions with the same friends, and contrary to popular opinion, probably enhances relationships. But it should not be permitted to interfere with other activities.

You play a part, too. Remember that all children look to parent behavior as a main way to learn social interactions. Maintaining good relationships with your children is more important than your own convenience. When interacting with your baby, turn off the TV or radio and be fully present. Otherwise, your distractions will weaken your interactions.

Avoid listening to your MP3 player with earphones, or speaking excessively on your cell phone and shushing your child. You are ignoring him and modeling antisocial use of media, not to mention missing out on a chance to interact. If you do make such a mistake, apologize immediately and take the opportunity to teach etiquette. Teach your children to be mindful so that they are not inadvertently rude, causing others to feel ignored when their phone rings and they answer it loudly. Talk together about how to decide and compromise between their urge to talk and the needs of others.

Socialization is pretty much an essential, natural, and healthy part of most non-solitary media activities. It is so pervasive, that pulling it apart from other Growth Opportunities and giving maximum weekly times is somewhat arbitrary and unrealistic. In recommending times for Socialization here, I am considering mainly of the negatives: Using interactive media to substitute for otherwise easily accessible in-person interactions in the absence of any other Growth Opportunity. An example would be texting a friend who can easily be met in person, or messaging trivia impulsively and thoughtlessly.

Interactive media can provide such immediate accessibility on a mere whim or push of a button, that children may not learn to delay gratification and develop patience and the skills they will need later to cope with life's frustrations. It is important for children to learn to tolerate separation from parents. They need to learn to wait and think through the whys and hows of interacting with others. In addition, learning to live in a sea of information requires sensitivity to privacy and the right to freedom from intrusion, triviality, exploitation, and information overload

For these reasons, I recommend less than an hour a week by age eight for this Growth Opportunity.

VALUES EDUCATION

Values Education is closely related to Family Relationships and Socialization, especially in the younger child. No matter what other Growth Opportunities interactive media provide, how they are used also presents opportunities for Values Education. Almost all human interactions call for consideration of others and/or ethical judgment. Preparing our children for life in a complex society that requires value judgments is a crucial GO, too often overlooked.

We need values to guide us. The Golden Rule is the basis for any value system. We try not to do something thoughtlessly, especially to others, just because we want to. We know what is good and do not remain indifferent in the face of what's bad.

Conversely, interactive media can weaken values unless used properly. Visiting the site of a hate group can provide an opportunity to teach critical thinking, discuss prejudice, injustice, and man's inhumanity to man. Violent games and destructive musical lyrics can cause increased aggression and poor school performance. Terrorists and other fringe groups recruit and teach their hate and craft online. Online harassment and bullying signal moral and ethical deficits in the perpetrators. Bullying is an example of children's cruelty, disrespect, and failure of major institutions, including the family. These misuses of interactive media can actually reverse the benefits of other parts of a Media Plan, especially for younger children.

I believe there is an urgent need for interactive media to provide values-oriented content. As a doctor and teacher, I am frequently asked by parents some sort of a "What shall I do?" question. Sometimes I can give a clear answer. Often, however, the only correct answer is, "It is really who you are that matters most to your kids. What you do comes from who you are. You must be alert, active, and fully present with your child. You must behave ethically and kindly. Your children are always watching and learning."

I cannot overemphasize this: How you interact with your child and how you treat him and others, including other family members, through or around media, is often much more important than any specific content you select. How you model your own values, how much respect, kindness, consideration, and patience you show him and others, and how you demonstrate the power of gratitude and forgiveness – that is how you teach your values.

I suggest that parents place Values Education high on their list of nutrients in children's Media Plans from an early age, and look for this Growth Opportunity in all activities. In addition, I urge parents to provide activities primarily on this crucial Growth Opportunity for at least thirty minutes weekly by age five and one hour by age eight and double or triple these times to include other media. This is one Growth Opportunity that you can give in mega-doses without any side effects.

EDUCATION ENRICHMENT

Education Enrichment is the main reason parents of very young children justify interactive media in their homes. But we don't really know for sure if in itself this expectation is warranted, since the subject is controversial, even among scholars. We need more long-term studies of the consequences of exposure to interactive media in infancy. However, in older kids, the initial unqualified excitement about a computer for every child has become significantly tempered. While once widely used, fewer parents are actually using educational software today.

Most scholars agree that how parents guide their children makes all the difference. So it is up to you. But realize that Baby is not an empty vessel into which you would pour whatever you will. She is an active learner on her own timetable. In her own way, and at her own pace, she selects what she needs to learn and when she needs it from the physical and emotional environment you give her. But she is delicate and can be harmed by too much stimulation.

Basing self-esteem on actual accomplishments rather than wishful thinking or the approval of others is essential for development and emotional health. Pleasure from mastery increases self-esteem, enhances the effectiveness of other nutrients, and makes for a healthy psychological life. Such pleasure can motivate continuing self improvement throughout life. Chances for increasing self-esteem are greatest when parent and child join in celebrating a child's new achievement. Even in adulthood, many of us continue to need parental approval. Eventually, if he has a good basis of receiving realistic approval at home, your child will require less outside approval for his success and will be able to provide himself with sufficient self-recognition.

Promote a child's feeling that he can successfully impact his world by working harder at what does not come easily to him. Coach your child to act in ways that give a sense of mastery while he remains generous and inclusive of others.

Stimulation of imagination and creativity enrich the child's inner experience of himself and make both internal and external discoveries possible. Creativity makes experimenting possible and allows him to apply what he has already learned in novel ways. Valuing his own imagination and striving to actively create, alone or collaboratively, are parts of active learning.

At every stage of development, encourage your child to use interactive media to extend and strengthen his own style of learning. Encourage your child to delve into areas where he has talents and interests, forming lifetime habits of self-directed learning and self-improvement. (Innovative and powerful learning environments are yet to be developed that combine interactivity, virtual reality and graphic representations, enormous reach, and access to vast resources and diverse communities – see Chapter 14.)

Children with special needs can benefit from the wealth of assistive software and online resources that interactive media offer. Make use of storage devices like e-readers, memory cards for PDAs, and flash drives that make accessible lots of visual and auditory information otherwise available only in bulky books from libraries.

Interactive media can enable more exciting, collaborative thinking as the child grows. Children can teach other children and learn from experts or work with others across the globe on commonly shared projects. Likewise, child and parent can enjoy the excitement of collaborative thinking and teach each other as they jointly interact with media. Innovative interactive devices like e-readers and search engines and databases provide instant access to books, periodicals, directories, and other interactive opportunities.

Learning to learn, teach, and collaborate start early when a parent celebrates each of Baby's new gains. Interactive media experiences can cause children to love learning new skills, mastering them, enjoying learning new things and proudly showing them off. In older children, parents should notice and reward self-discipline, originality, curiosity, perseverance, determination, focus and avoidance of distractions. Notice and reward gains in organization, integration, problem-solving, and retention. Be explicit in your praise when a child shows respect for others, values his own intellect, and interacts and cross fertilizes ideas with others.

Computer literacy is now an essential component of modern life. While children know how to operate computers and interact online, few have a good grasp of how computers and the Internet actually work. Beginning with first grade, some technical education suited to the child's age would lead to better appreciation of the potential, hazards and benefits of these interconnected machines. All family members can benefit from learning more and keeping up with new developments by subscribing to technology alerts offered by newspapers or other online sources.

Today, children consume and produce vast quantities of information and will continue to do so throughout their lives. All brains, especially babies', can only accept so much before they become overloaded. When that happens, children may become irritable or lose interest and space out. Parents can teach older kids how to use critical thinking skills to examine content and form of media for their benefit.

More than prior generations, our children will need skills in processing information, thinking critically and evaluating carefully to benefit from these oceans of data. With torrents of information flowing around us, we face so many more decisions about what to take in and what to ignore, what is reliable and what is misleading, what is science and what is superstition, and what is authentic and what is false.

Just as developing good eating habits is important, so is learning good interactive media practices to cope with information received and communicated digitally from so many sources so quickly. But there may be a downside here: While computer keyboarding is a skill taught and learned early, fine motor control and eye-hand coordination learned through cursive writing may suffer.

Education Enrichment starts in early infancy and continues through life. The content of such activity may be educational, but the interactivity may be social or with family members. For the purpose of quantifying time spent on this activity as a primary goal at home, I recommend a minimum of thirty minutes by age five and forty-five minutes by age eight. This is another activity to indulge children with.

ENTERTAINMENT

There are few experiences that do not bring Baby the wonders of new discoveries. In older kids, interactive media Entertainment activities can also enhance all other Growth Opportunities by enabling sportsmanship, cooperation, and collaboration. Playing toward a common goal with a parent brings the pleasure of collaboration. Games can teach motor skills, strategy, discipline, and focus, even for two-year-olds. These benefits may lead to the long-term acquisition of new skill sets.

Today, the video game is considered by many to constitute a serious art form that reflects popular culture, much as the best films and comic books do. Music scores for many games are as well-crafted and enjoyable as sound tracks in movies; and graphics, made possible by more powerful computers, are as enchanting as the best cartoons of old.

Developmentally appropriate play, in which the evolving child rehearses the potentials of his mind and body, is a crucial task and the most serious work of children. The child masters the old and invents the new. It is a way that children take in new experiences and work them out for themselves, growing and expanding their world in the process. Play brings pleasure from mastery and stimulates imagination and creativity. But play is much more than fun, just as a much more than important nutrients.

Fun is one appropriate response to learning in young children and can be an end in itself. Children seek to learn because it is fun. In older children, fun for its own sake is good in limited amounts. But, like spices, fun is best when blended as a powerful motivator for consumption of other nutrients. Fun is an essential ingredient for learning, as are vitamins in a food diet. When mixed in, fun or pleasure is a great motivator. Yet, fun is not an ingredient that can stand on its own, any more than paprika or cinnamon. While not an essential nutrient, fun can usually make the difference between a successful dish and a failure.

Fun is always important, but should not predominate. There is a significant potential downside to the abundance of opportunities offered in our culture by interactive media. Too many kids fill in every moment of their lives with activities—learning, socializing, and doing, doing, doing ... They complain of boredom if not always busy. Kids need time for quiet play to think, explore, invent, solve problems, and otherwise experiment with ideas and roles in their own ways. Kids need downtime to integrate all else that they experience through original, creative thoughts and insights that are not merely reactions to someone else's. They need to be present in their own lives without boredom.

The Media Plan will guide you in how much Entertainment is best for your child. I hope that commercial entertainment interactive media will eventually include more Values Education and opportunities for Family Relationships. There are encouraging applications of video games in education. In general, however, there does not seem to be any compelling reason for young schoolchildren to have free access to Entertainment either at home or via mobile devices or cell phones.

Bottom line: These essential ingredients together impact the child and carry him toward becoming a smoothly functioning and integrated person. Just as food helps a child to grow, so must interactive media promote development and elicit appropriate responses from the child. Look at your child's face to see how engaged he is. The quality, novelty, and form of an experience determine the amount and type of interactivity. For example, if a site evokes curiosity and excitement about a topic, it is appropriate. If it evokes boredom or provokes aggression, it is not. The rate of interaction and stimulation is important in holding and sustaining attention and in keeping young children interested.

Chapter 3: Setup and Installation of the Media Plan

When it comes to preparing meals, every parent usually knows what to do and is careful to provide good nourishment to protect the family from unsafe or harmful foods. You know that there are harmful foods, but there are also good and necessary foods as well as fun foods. Knowing, finding, buying and preparing content in a correct balance, and presenting it in ways that are appropriate for your child is an ongoing challenge to parents, just like properly feeding the family.

Do you design the same menu and use the same eating utensils for an infant as for a seven-year-old? Of course not. Likewise, each child evolves differently and needs new mixes of interactive media as he grows older. A Media Plan provides a balance of essential interactive media nutrients in the right proportions for the child's age. Your choices are more limited for infants than for older children, both in food and in media experiences. But they are just as crucial, if not more so.

A reasonably functioning family should be able to follow my guidelines. Such a family is physically and emotionally safe and reasonably nurturing and supportive to all its members. It supports the unique developmental trajectory of its adults, as well as of children. The child's life is free of unusual stressors. The child functions reasonably well at home, day care, or school academically, socially, and morally. The child's life is not crammed with activities. The child has sufficient time, free of electronics or social stimulation, to learn to be alone and discover aspects of his world for himself, without immediately complaining of being bored. If this does not describe your family in large part, consider it a wake-up call and consider making changes.

Now is the time to go the final distance. Earlier, you began to understand the place of interactive media in your family's life and learned about the Media Plan for creating the right mix of Growth Opportunities. Now you are ready to equip yourself, find the proper mix of online sites and other media activities for your child, create your actual Media Plan, and implement the plan into your daily life. In this chapter, you will come closer to incorporating a framework for thinking and planning the media experience for your children starting in infancy and continuing throughout childhood.

RECOMMENDATIONS FOR THE MEDIA PLAN BY AGE

My guidelines are merely thoughtful suggestions. As a sensitive partner to your child in promoting his development, you really are the expert.

The times on the charts and tables below reflect general recommendations for what is best for children. Time spent with interactive media increases with maturation. Also, as the child gets older, the total time for a parent's direct supervision or involvement decreases, while independent media time increases.

Of course, it is unrealistic to regiment children too rigidly by specifying precise times for all media activities, so please use your judgment. For example, add more Socialization or Entertainment time as rewards. Additionally, try to find the balance between control and permissiveness that works for you and the child. Remember that self-directed learning and freedom to explore are important aspects of development. Give him and yourself enough room to make mistakes. With your older child, review the plan on a regular basis.

Be sure to always keep in mind safety, protection from media exploitation, and other principles you learn here. Treat your child and other family members with patience, respect, and kindness. Do not allow interactive media activities to interfere with meals, bedtimes, and other family routines.

Customize the times for a child with unusual or special needs, or according to your own preferences. Add more Socialization or Entertainment time as rewards. The proportional balance among Growth Opportunities is as important as the amount of time spent.

Remain consistent and make changes systematically. As a general rule, to include TV, movies, viewing sports, and music, you may double or triple the times given in the table. Obviously, the recommendation for maximum length per session cannot always apply. It is best to stick to the relative proportions of Growth Opportunities given for digital media. I do not include times reading online or elsewhere as digital media. An excess of any activity is usually unhealthy.

Here's how to use the guideline table below: To decide actual times and proportions right for your child, first determine his age and then find his approximate place within the table. Calculate his actual times within the range given. The graphical illustrations that follow provide the same information in a quick visual form. If a child's maturity lags or is ahead of his age group, select the appropriate age in the table.

Looking across the top row, first find your child's age. For example, if your child is five, run your eyes down the center column to see the weekly time guidelines for the Growth Opportunities and other Media Plan aspects listed in the column on the very left. You will note that the total maximum weekly time would be three hours, with a maximum of 45 minutes a session and almost three hours with your active participation. You may decide on times that fit your child's level of maturity and interests.

For ages that fall between those listed, estimate the approximate times. For example, if your child is seven, the guidelines would be approximately halfway between those for five and eight.

Times apply to the total amount to be spent on all interactive digital media, excluding school work. I also provide rough suggested maximum times for other, non-digital. media, such as movies, concerts, sports events, and TV. These are not allocated in the tables.

Start with the times I recommend, but change them as you gradually discover the best arrangements for your family. Or start with what you are already doing and work toward these goals. One single media activity can provide several Growth Opportunities at the same time. For example, playing a video game with a friend for an hour that provides a strong educational component can count at the same time as an hour of Socialization, an hour of Education Enrichment, and an hour of Entertainment.

MEDIA PLAN GUIDELINES BY AGE			
Hours /Week (Digital)	Age 2	Age 5	Age 8
minimum Family Relations	1:00	2:00	2:30
maximum Socialization			:30
minimum Values Education		:30	1:00
minimum Education Enrichment		:30	:45
maximum Entertainment			:15
total maximum	1:00	3:00	5:00
total minimum			1:00
maximum per session	:15	:45	1:00
minimum with parent	1:00	2:45	2:45
minimum parent nearby		:15	2:00
maximum independent			:15
+ Hours /Week Other Media	3:00	6:00	10:00

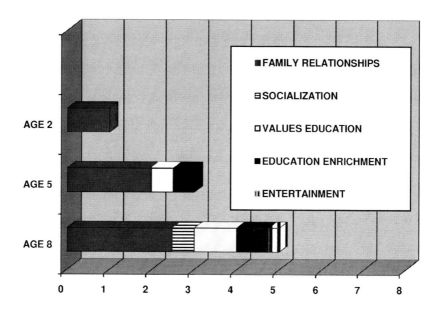

Figure 3.1: For ages 2 through 8, here are guidelines for weekly hours of consumption of digital media.

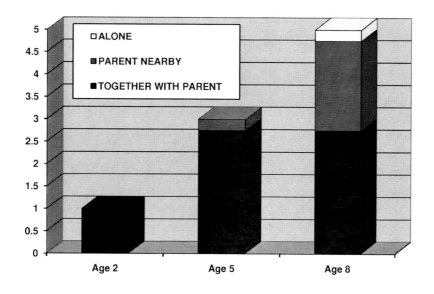

Figure 3.2: For ages 2 through 8, here are guidelines for weekly hours of parent presence while the child is interacting with digital media.

THE HOME SPACE

What we want to do in a room determines how we arrange our furniture. In turn, how we place the furniture encourages or discourages how we use a room. What kind of space would promote interpersonal interactions around the interactive family computer?

Altering the physical arrangement can encourage people to look at each other and enhance social and Family Relationships. While putting several chairs in front of the screen can allow for important interpersonal experiences, available products are just beginning to enable more complex interpersonal interactions. In this rapidly evolving field, new systems are introduced at a rapid pace. For example, Windows XP can accommodate several HIDs (Human Interface Devices like wireless keyboards and mice) simultaneously via USB plug-ins. The technology is already here—it just needs assembling.

STEP-BY-STEP

SEARCH WORDS: child, software, safe, hardware, learning, family

Like any good cook, you need a kitchen equipped with the right equipment, utensils, and groceries. Similarly, you will use each online and interactive media experience to contribute substantially to the emotional, social, motor, cognitive, moral, self care skills and aesthetic development of your child through a proper mix of Growth Opportunities. Just as some children require special food diets, some may require special Media Plans. Plan for your youngest child first, or whoever is easiest, and choose how far you want to go.

Basic hardware and seating. Think of hardware and software accessories that will help you accomplish your present goals and will also set you up for the future. These may include sound, graphics, touch screen displays, and webcam communications capabilities for the young child.

If have not done so already, I suggest strongly that you get familiar with roaming online, following hyperlinks, and using e-mail if you have not done so already. Set up e-mail accounts for each family member and send each other test messages. Use your computer to keep track of your efforts and notes as well as to share with others. Parents may copy the forms and tables in this book for their home use or see them online at www.mydigitalfamily.org. . A fact little known by merchants, and even manufacturers' help desks, a computer is that some operating systems (for example, Windows XP) can accept several keyboards and mice.

Start investigating and pricing resources in stores or from online merchants. Appropriate hardware and computer accessories and their helpful arrangement are important in changing from solitary computer use to enabling the online family concept.

Educate yourself about available educational and filtering software. As your menu ideas get clearer, you may need to equip yourself further.

Set up a special physical space to allow a comfortable new psychological space in your home. Create the physical arrangement that can enhance family cooperation, and verbal and non-verbal communication through being online, playing, and learning together. Figure out how you will add seating. You might even move the computer to a place where everybody can use it. Remember, you are making a new physical space to accommodate a new and important family activity. Setting up itself can work as a family activity. The family room may be a good place, and some video displays can serve as both TVs and computer monitors.

Be creative. Some merchants may be willing put a package together for you, or you may be able to design a package for yourself. Eventually, merchants might bundle of hardware, furniture, and software that enable family computing.

How many computers? Families have children with differing interests in different age groups contending for interactive media access at the same time. This challenge for parents is familiar: how can I allocate resources, while developing and scheduling activities so that everyone has an appropriate experience? The answer is, "Usually pretty well, but not perfectly." You already know how to do it. After all, one of the advantages of having brothers and sisters is learning how to live and share with others.

Siblings who have needs in common can share the same online time. Sessions at the single online family computer may have to be segmented, with shared times and alone times. For instance, all can share family together time on a virtual tour of the White House or Alaska. However, the older child may wish privacy in instant messaging his friends. In some instances, you may have to network more than one online computer for homework or Education Enrichment, with one online family computer and several work stations. Or a young child may download his homework materials via the online family computer and transfer it to his work station in his room via the home network, flash drive, or CD.

The trend to include Internet access by mobile devices is universal. I see no reason for grade-school children to have such access routinely, and advise parents to block it via a simple call to the provider.

Gather information. Give yourself at least one month to build your network and gather information. Start at how your child uses the family computer already. Use the Media Plan Worksheet on page 44 to estimate how much time he spends online and on what activities. Having the family computer within range will enable you to make these estimates more accurately. For example, if an older child can cooperate honestly, ask her to keep the log for a few days, or work together to make the estimates. Keep current with rapidly evolving interactive media. Ask your children to contribute suggestions. Develop a network of resources and advocate anywhere you can for the development of more resources.

Work on a Media Plan for one child at a time. Your goal is to develop a working list of at least twenty Internet sites suitable to your child for each Growth Opportunity. You will need to use critical judgment. For example, doctors from St. Michael's Hospital in Toronto found that they could recommend only fourteen of the fifty-five sites they found on safety education. You will later assemble sites from this list into the Media Plan for your child, and allocate them to daily sessions, each with an appropriate menu. An older sibling may enjoy helping you design a Media Plan for her younger sister or brother.

Media Plan goals can be achieved in many ways, and there is no limit to the creativity of the menu planner in arranging interactive media experiences suitable to the needs of the child. Try hard to discover, select, learn, and analyze the information available to find just the right interactive media experiences for your child. See *www.mydigitalfamily.org* for ideas and help. As you would with a meal, plan online sessions with the correct balance of essential media nutrients grouped in Growth Opportunities that promote healthy functioning.

Ask other family members what they have visited, enjoyed, and found useful online. Include the child's teacher(s) in your planning—especially for the young or special needs child. An additional superb resource is your school librarian or local community librarian. Many of these professionals have become experts in the software and online learning and enrichment to children, as well the pitfalls and hazards.

These individuals often have organized resources to help parents negotiate the world of interactive media. Don't forget to ask your friends and other parents. In fact, you may want to get a group of parents together to discuss these issues regularly. This could include a mother's group, play group, PTA or church committee, or another forum. You may want to start an e-mail newsletter, social site group (one way for you to get firsthand experience in what your kids are so good at), or even a Web site aimed at implementing Media Plans with sample plans and menus. In this way you may also become more proficient in the use of the computer for communications and joint exploration, skills you may teach your young children (your older children probably already know these, even better than you, and can guide you).

ASK TEACHERS AND OTHERS ABOUT YOUR CHILD'S NEEDS

- How do you see my child's specific needs and strengths?
- Does my child have any talents to develop?
- What Internet sites or other interactive media would be good for a child my son or daughter's age?
- Help my child use media to meet her needs, work with her deficits, and enhance her strengths?
- What are my child's specific strengths and needs in these developmental areas: sensorimotor, cognitive, emotional, social, moral, self-care skills, and aesthetics?

> - Can you think of educational (or health, social, spiritual, social, etc.) computer material that would enrich my child and supplement what you are already doing (in school, in church, etc.)?

Harvest opportunities using the Internet and other interactive media. As you research resources online, look for a variety and be creative in fitting what catches your eye into your overall plan. Look at www.mydigitalfamily.org for help. Use search engines to discover a variety of relevant or helpful online sites. You will learn quickly how to find what you are looking for on the Internet. Ask your older child to join you. A site may provide more than one Growth Opportunity.

Look for online sites regularly updated by education groups. Make your favorite search site your home page. Use any word that will lead you to a special interest, in combination with "children" and a developmental goal or nutrient. Follow links provided in search engines' listings and then links provided by the various sites.

Build consensus. As you work online together, accommodate your needs along with your child's needs and preferences.

Use the Media Plan Worksheet on page 44 to organize and group sites by Growth Opportunities. It can assist you when presenting your findings to your family. Your name atop the Worksheet lends it authority and personalizes the project. Sign and date the forms with all family members, including your child, to record her agreement, empower her, and make the whole process more official. This activity in itself can introduce the concept of a contract.

Aim to end up with at least twenty sites grouped by the Growth Opportunities they provide. Copy links into your browser favorites or your word processor and keep track with notes. Later, you can transfer these links to the child's media portal.

WHAT TO LOOK FOR ONLINE AND OTHER MEDIA
SEARCH WORDS: family portal Internet home

NOURISHMENT FOR KIDS
- Family Relationships (activities for kids and parents together)
- Socialization (social development)
- Values Education (moral development)
- Education Enrichment (cognitive development)
- Entertainment
- Sensorimotor development
- Emotional development

- Self-care skills development
- Aesthetics opportunities for development
- Activities for kids, some by age groups
- Search engines for kids
- Collaborative thinking and cooperation with others

SUPPORT FOR PARENTS
- Resources for parents (mostly safety-oriented materials)
- Family friendly content (most are OK, but be selective)
- Family friendly online directories, some sponsored by search engine companies
- Commercial services and products—internet service providers (ISPs) and software that provide levels of filtering appropriate to children's ages.
- Sites and directories for special groups; for example, religious or ethnic
- Information about Internet use, crime, ethical media marketing and trends

LEAD YOUR FAMILY

There cannot be a good and safe family without good parental leadership. The Media Plan needs a functional family. There can be no system without it. Such families are not perfect, but they tend to resolve conflicts without excessive tension. Functional families problem solve together; take opportunities to enjoy and strengthen appropriate roles; provide many more positive than negative experiences, and practice mutual respect for boundaries. Mutual trust and a good family life are crucial for safe and healthy interactive media experiences. Knowing the needs of your child and treating her with patience, consideration, and respect are keys to the success of your plan. Family members take responsibility for their actions and acknowledge its effects on others, apologizing when wrong. Remember: "Leaders aren't born, they are made. And they are made just like everything else, through hard work. And that's the price we'll have to pay to achieve … any goal" (Vincent Lombardi).

EXAMPLE

Grandfather was coming over Friday for dinner and wanted to be with the children. So Mother scheduled him to go online together with Jana, eight, her six year old brother, Robert, and her sleepover friend, Maria, seven, between 5:30 pm and 7:00 pm. Accordingly, Mom and the kids decided to use Jana's Media Portal Page and prepared it ahead of time.

At about 5:30, Jana, Robert, Maria, and Grandfather all sat around the computer as they went online to visit whitehouse.gov. They interacted face-to- face and enjoyed the activity together.

Comment: In this way, both Jana's and Robert's Media Plan Growth Opportunity requirements for Family Relations, Socialization, and Education Enrichment and their Parent Present guidelines were all satisfied at the same time (but would be added as just 1/2 hour to their Total Media Maximums. Additionally, because Maria herself is also on a Media Plan in her own home, Jana's mom let Maria's mom know exactly how Maria spent her media time during her sleepover with Jana.

CHILDREN WITH SPECIAL NEEDS

SEARCH WORDS: Internet computer support special need child special education

So far we have seen how "average" children can benefit from a Media Plan. But, clearly, "average" is a myth and only a theoretical notion because each child is unique. Every baby has a unique temperament. Media exposure must suit the baby's temperament. Some babies crave the stimulation of touch and movement, while others are so sensitive that they find even normal touch or ordinary sounds unpleasant.

Thoughtfully assessing (and regularly reassessing) the unique needs of every child, especially of a special needs child, is always a good idea. Many children require Media Plans outside the general range, especially those with special needs.

We would consider any child requiring extra consideration or special education services as having special needs. These would include children with chronic illnesses, neurological deficits, diabetes, and severe obesity. Also, those who suffer from hyperactivity and impulse- or attention- deficits, psychological trauma, depression, anxiety, over-attachment, social deficits and shyness; intellectual, learning, and language deficits; over-aggressive, antisocial, or substance-abusing children, as well as the bully and the bullied child; and those suffering from motor and sensory deficits - these are all examples of special needs children. Additionally, children living in poor, minority, extremely wealthy, immigrant or bilingual families, foster children, or children in hospitals or special schools all have their own special needs.

The gifted child has special needs too, because she is ahead of same-age peers in one or more areas of development: sensorimotor (athletics, dancing, painting, sculpture, ceramics), cognitive (high IQ or special language, writing, or math abilities), emotional (exceptionally sensitive and responsive to other's needs without giving up own personal strengths), social (especially adept at social relationships), moral (having a highly developed sense of right and wrong and ready to act on behalf of this sense), and aesthetics (highly developed musicality or other artistic talent). Many times the talent is so great that it dwarfs other areas.

Dr. Eitan D. Schwarz

MEDIA PLAN WORKSHEET For_____

TOTAL ONLINE TIME / WEEK__HRS   MAX__HR / SESSION

ALONE_____hr PARENT NEARBY____hr   WITH PARENT____hr

FAMILY RELATIONSHIPS

SOCIALIZATION

VALUES EDUCATION

EDUCATION ENRICHMENT

ENTERTAINMENT

NON-DIGITAL MEDIA

We pledge as a family to always show respect, kindness, patience and to use the computer safely. We all agree not to snack in front of the computer and not to let online activities interfere with bedtime, mealtime, or other times. We pledge to help each other keep this schedule.

ALL REVIEWED AND AGREED TO SAFETY RULES

Signed _____ _____ _____ _____ Date _____

MAKING THE DIGITAL MEDIA PORTAL PAGE

- Find media resources and complete the Media Plan Worksheet.
- Get a general idea for your child of the proportions of Growth Opportunities and your suggested presence from the table and graphs on page 36-37. Estimate approximate daily times for each Growth Opportunity.
- Make a Media Portal Page on a word processor or personal web page (that you and your child can decorate).
- Paste/copy links from the Media Plan Worksheet on page 44 onto the Media Portal Page.
- This page can serve as the child's gateway for all digital media activities.
- Include in the estimates for TV, video game, music, phone and texting times and times spent with mobile devices.
- Print and post the Media Portal Page.
- Teach and ask the child to participate in an honor system using the computer only through this Media Portal Page.
- Start child off with the Media Portal Page on screen and keep track roughly of times spent on each Growth Opportunity and your required presence.
- Teach telling time or use a mechanical or electronic or computer timer.
- Schedule yourself to be present and involved, as suggested by the guidelines.
- To make more detailed daily menus, divide up weekly total times in the table on page 36 among days of the week, trying similarly to allocate the other guidelines.
- Review and revise the Media Portal Page regularly.

Parents of special needs and gifted children must give extra thought to formulating the content of their Media Plan and its level of intensity. Theirs can start as a general plan, but may be modified to provide extra amounts of essential online and interactive media nutrients to support maturation or development in specific areas. Additionally, some essential online and interactive media nutrients may be reduced or removed to meet the child's needs. The child may require essential ingredients balanced specially or in a form more readily accessible.

Start by looking at your child with fresh eyes. Be realistic and honest with yourself. The more thorough your efforts to discover how things really are now, the better you will be able to decide how you want them to be later. The first step is to think through what you and other family members know about your child. List the developmental tasks of childhood. Describe your child under these headings and try to explore where he is now and where he is heading. If possible, brainstorming is a great idea at this stage. You can review the appropriate chapter for more detail.

Consult closely with your child's pediatrician, developmental expert, teachers, and professional helpers during the information-gathering stage. Set up appointments of approximately an hour in length with key caretakers and teachers, and interview them thoroughly especially for this purpose.

One area that has become quite well-developed already is the use of the on- and offline computer in schools, especially in special education programs. For instance, sophisticated interactive media can support a dyslexic child's learning needs by transcribing the child's spoken words into printed text. Your child's school may already include an interactive media assessment as part of the special education assessment procedures to discover what media can assist your child. Ask for it if it is not offered. Often, ideas from your child's existing school plan can help you generate ideas for your Media Plan.

By now you will have a pretty good idea of your child's strengths, preferences, and areas needing special work. Putting it all together should be easier since you created categories for the information that focus on your child's developmental needs. Now you are ready to set goals and make a plan. Summarize the child's accomplishments and tasks in each developmental area (sensorimotor and cognitive, emotional, social, moral, self care skills, and aesthetic), and set basic goals in each. Your plan would provide essential online and interactive media nutrients and Growth Opportunities in suitable proportions and forms.

In general, to set goals for your child's special needs, you will seek balance among optimizing development, building strengths, rehabilitating deficits, maintaining hope and self-esteem, as well as avoiding frustration and failure. To achieve these goals, you might need to consider accommodations in the physical space, input and output devices, software, and online sites.

You may want to start or join a chat group to investigate Internet and other interactive media resources. Contact other parents in similar circumstances, reputable self-help groups, and national organizations when working with your special needs child. Brainstorm and obtain advice and guidance. Beware, however, of unsubstantiated claims and testimonials. Advocate for more attention to the online needs of children by these groups, organizations, and experts.

FORWARD TO THE FUTURE
OUR FAMILYSITE.FAM

Consider building a password-protected family site or social media page together for posting photos, news, announcements, and other family information. Include distant relatives in the project. Provide chat room, webcam, or texting functions if possible. Teach ethical and safe online habits by posting safety and online behavior rules and consequences for violating them, as well as news in these areas. Have each child design his own page and post photos, writings, and scanned drawings.

People with similar interests can have a forum of their own; for instance: knitting, cooking, traveling, hunting, or shopping. Large families can reserve a page for relatives of each age group. A child together with a grown-up can partner as webmasters, and everybody can take turns maintaining and expanding the site. Reserve for each child a page she may decorate herself that also contains her Media Plan Worksheet with schedules and links to permitted sites.

The possibilities are endless and will only grow as you continue to grow along with technology.

HOW INVOLVED WILL YOU GET?

Choose the level of involvement best for you.

BASIC INTEREST— A GOOD WAY TO GET STARTED: Commit yourself to being present in your child's media life. Think about how the principles you learned can be applied to other areas of parenting.

Make sure the computer is always within earshot. Limit media time. Assure safety and filter out unsafe sites. Accept that your child might use interactive media relatively unsupervised and primarily for Entertainment and/or Socialization, but introduce your family to the idea of the Media Plan. For example, decide together how much total weekly time the child will be online and playing video games after completing homework. Keep an eye on your child and monitor his media activities regularly, including use of mobile devices.

Pull up a chair and join the child occasionally and look for sites that provide Values Education, Family Relationships, and Education Enrichment. Make media use a family activity. Use the Media Plan Worksheet on page 44 to organize your thoughts and notes. Review safety, monitor progress. Readjust your plan occasionally. Be consistent and follow through the best you can.

ACTIVELY INVOLVED: Build on the involvement described above. Keep several chairs around the family computer. Decide on media goals for your child and family. Tell your child the types of sites you permit and the types you encourage. Build family consensus and empower your child. Broach the subject of media marketing to shield or inoculate your child. Sign the Media Plan Worksheet on page 44 together with your child.

INFORMAL: Build on the levels described above. Use the Media Plan Worksheet on page 44 as a log to estimate how much time your child actually spends on each Growth Opportunity. Unless there are compelling reasons for their use, consider calling the cell phone provider to block Internet access. Readjust your plan occasionally.

Form a basic plan based on your child's age and needs using the guidelines in this book. Look for sites and other activities that meet your general goals and your child's needs. List sites and/or times per week for each Growth Opportunity on the Media Plan Worksheet. Create a Media Portal Page, outlined on page 45.

For more rigorous involvements, please see User's Guide for MAKE A CASUAL and GOURMET methods (page 128.)

Chapter 4: Keeping Your Baby's Brain "Green" (Birth to Two)

Babies are so different from us, not only in the way they look, but even more in the ways they see, hear, eat, smell, and otherwise sense and act on the world. They might as well be aliens from another planet. They are so exotic that popular science fiction often portrays extraterrestrials as resembling human babies. Science fiction often also picks up on how quickly babies soak up knowledge as they attempt to learn about the human race.

Although relatively little is known about how babies and toddlers interact with newer media, over the past several decades we have already learned a lot from the research about the impact of television screen time. It seems that TV is not at all a passive medium for babies. Watch them. When they watch, they really watch. They are actually very actively engaged and have been discovered by scientists to react psychologically and emotionally.

I know that parents don't have it easy and must largely chart their own course regarding use of media. Contrary to pop wisdom, we really don't know that babies can be hurried into becoming Mozarts or Einsteins. I do believe, however, that parents can introduce rich areas of positive interactions into relationships with their young children and lay a solid foundation for media literacy and cooperative use of interactive media and intelligent consumerism.

There is no hurry. By starting the Media Plan early, however, you can probably prevent future power struggles and addiction (Interactive Media Overuse Syndrome). If you teach your child from the beginning to know interactive media as family-centered tools for much more than Entertainment, these lessons will pay dividends as she grows. But you must always be fully present, because, in my view, the main reason for Baby to be exposed to interactive media and screen time is to strengthen your relationship with your very young child.

BABY'S BRAIN IS A VERY SPECIAL ENVIRONMENT

Aside from basic physical survival skills, the mature human brain with its enormous complexity and computational power is primarily designed to enable survival through social/emotional/intellectual connections with fellow humans. Baby possesses her amazing brain, in all its wonderful plasticity, primarily as a means to develop a more mature brain that can eventually process the information she will need for living among her fellow human creatures.

In evolutionary terms, our human brains is still a work in progress—we have a long way to go in managing aggression and living together more successfully. Nevertheless, I believe that everything else man has accomplished during his relatively brief time as a civilized being, including all his awesome intellectual, spiritual, and aesthetic accomplishments, is secondary to this one grand purpose of interacting with each other. If nothing else, these wonderful talents and accomplishments enrich our interactions with one another and show just how much power the human brain must have to accomplish its one main task of managing our social world.

Baby's immature brain is vastly different from an older child's in how it looks under the microscope and how it works. The more scientists learn about a baby's brain, the more they are awed by one of the most amazing miracles in nature; not only in its basic design, but how it is programmed to build itself in response to its surroundings.

Baby's brain must eventually develop into a resource that enables her to succeed in a complex world. Her brain must become the control center for choreographing ever-changing patterns of appropriate behaviors that result from good judgment. She will eventually need to fully form a mind, the integrated sum total of her brain's activities, including what makes her human—consciousness, self-awareness, creativity, and the ability to direct her attention at will. She will need to form judgments from accurate perceptions that lead her to achieve wise balances and compromises of urges, thoughts, and feelings. Her brain will not be this fully mature until she is in her early twenties. She has a long way to go. Supporting her brain's maturation and her mind's development is the main task of her family, community, and an increasingly complex web of societies. The more complex her society, the more preparation she will require.

Needing human contact to power her development, Baby comes preprogrammed. The younger the baby, the more dramatically she seems to accumulate daily gains, to the delight of parents charged with raising and protecting the future of mankind. No matter what her genetic inheritance, loving human interactions, sensitive touch, face-to-face talking, and other intimate contacts, are the most powerful organizers and stimulants for her brain's development and the foundation for everything else that is to follow.

Baby begins to show her inborn temperament early and transforms herself amazingly quickly into a fully interactional social/emotional/intellectual being, blending into herself aspects of the people and perceptions around her. Baby is not a passive recipient of new knowledge. She actively and naturally seeks the experiences that will transform her brain on her own schedule. Developing basic language and nonverbal communication skills is one primary task she must accomplish, and she is already well prewired to do so.

Better safe than sorry. These days we are super-aware of the delicacy of the world's environment and how we must be careful to keep it "green." The fuels that had energized the growth of our great nation last century now prove to have unintended side effects. We have learned increasingly to appreciate nature's delicate ecological balances and our responsibility to refrain from disturbing them. Think of this: babies are our most precious natural resource. A baby's brain is one of the most amazing ecologies in the universe in its biological complexity and how its potential is so exquisitely sensitive to our actions. It is an environment that requires from us our best greater, sensitivity, thoughtfulness, respect, and care. Baby needs us to keep her "green."

FORMULA OR BREAST MILK?

After abandoning it in the 1950s for a flirtation with manufactured baby formula, mothers themselves led us back to the natural art and joy of breast-feeding. We have subsequently come to appreciate that breast-feeding is far superior to formula for both baby and mother. When we try to improve on one of Mother Nature's most amazing designs, we ought to make darn sure we know what we are doing.

Exposure to media as we know it now is no substitute for human interaction, especially for babies and toddlers. Media is like baby formula. Screen time has had no special place in nature's grand design. We must remember all of this when we extrapolate from what we know about older children. We must remember this when we introduce changes without knowing fully if we will accomplish their intended effects without unanticipated harm.

Doctors usually make decisions by weighing the benefit vs. harm. For example, your doctor would recommend taking a chance even if the benefit may be small or uncertain (like in cancer chemotherapy). Or if the likely benefit is large while the risk is small, then it is even easier to provide guidance. Doctors try to learn as much as they can, but often, in this imperfect world, must make recommendations with only partial knowledge. So in 1999, taking all that was scientifically known into account, the prestigious American Academy of Pediatrics recommended that if Baby is younger than two, she should not be exposed to television.

The baby doctors knew enough to say that the benefits of TV didn't seem to justify its risks. The doctors assumed that their weighty credibility would influence parents. But they miscalculated. They were stunned that their guidelines were largely ignored, as have been years of similar admonitions by professionals and scholars backed by solid science. Although many parents are uncomfortable and report feeling embarrassed and guilty about putting children in front of a screen, they continue to do so in great numbers. Most children, even the very young, are watching lots of TV. Busy parents, themselves raised on TV, have come to rely heavily on TV and baby videos, at least to fill in periods of their children's day, or to allow themselves some downtime to take a shower. Some say that they just cannot manage otherwise in their own busy and demanding lives. Besides, wanting the best for their child, parents do not want to miss possible opportunities to help their development. They don't see the point of waiting for science to catch up with an aggressive commerce that floods them with powerful skilled marketing backed by paid "experts."

I want to make a point: Although claims by merchandisers often might as well be claims for the magic healing of colored water, many parents have been convinced that they are helping their babies learn by exposing them to TV and baby videos. Why? Because of the *Wizard of Oz* effect in human nature: Many parents would sometimes actually rather believe a fast-talking wizard engagingly hawking magic, healing, colored water from the back of a truck than a boring scientist. They easily believe that the wizard can make their baby brave like the Lion, smart like the Scarecrow, and kind like the Tin Man. The wizard would get baby to where she ought to be, just like he helped Dorothy with her magic, sparkly, red slippers.

The pediatricians, good doctors advocating for sound development, simply did not appreciate the power of cultural forces driven mainly by the modern wizards of merchandising. They did not pay enough attention to the zeal of well-meaning parents who would often overlook the fact that the only beneficiary is the guy proselytizing from the back of the truck, and that—worse—colored water can be accidentally harmful. Recently, a large media company ended up refunding customers for a promise that was not delivered by the Baby Einstein videos.

Bottom line: these days, parents are mostly left to themselves to manage their kids' interactions with media. Lack of knowledge and controversies among scientists, child advocates, educators, government, broadcasters, and merchandisers are confusing and still give off more heat than light. It is no wonder that the authors of two excellent recent summaries—*Buy, Buy Baby* and *Into the Minds of Babes*—are themselves concerned mothers driven to figure out for themselves what is actually going on. Along with many experts, these mothers are urging us to not repeat the earlier mistake we made in preferring baby formula to breast-feeding. I pick up where the author-mothers left off to give balanced, practical, and sound guidelines to parents.

Incidentally, one silver lining to all this hoopla has been the increasing appreciation of the importance of the first years of life to later development. Hopefully, public funding for programs supporting families, early education, research, and early abuse and violence prevention will increase.

EXAMPLE

Two year old Kara is cute and on track developmentally. Mother and daughter have wonderful loving moments they call their special "huggle" times. Kara facing mom and hugging, cuddling, playing patty-cake, singing and tickling punctuate the play sessions. These joyful moments often include Kara's "Teddy," and dad joins in if he happens to be home.

Mom does not answer the phone and makes it a priority to be fully present for uninterrupted play that provides opportunities for "huggle" times when the other kids are at school. Despite a busy schedule and a part-time job, she is able to fit in several such opportunities each week. When this is impossible, she asks dad to take out the other kids on the weekend.

Mom and Kara enjoy spending time together, looking at picture books, and singing. Kara is temperamentally active and can be stubborn. When Kara gets squirmy, she slips off mom's lap. Mom answers her when she calls from the next room, and accepts Kara back on her lap when she returns. Recently, mom put Kara on her lap while doing computer chores, positioning her to half face the screen, so that the two could easily keep eye contact, while at the same time seeing the screen. Kara started touching the keyboard and screen, and mom guided Kara's finger to push keys, sometimes naming a letter or number or color.

Kara started demanding mom's full attention, and without it would either slip away sullenly or become a nuisance. So mom now schedules computer chore time separately. She finds graphic Web sites that Kara likes and has given special names, and Kara often asks for the music software that produces sounds and colors. Occasionally, mom asks Kara if she wants a "special huggle" and helps Kara turn to hug, cuddle, and be tickled.

Kara is sometimes so engrossed by the responsiveness and stimulation of the computer that she does not care for a "special huggle." That reminds mom of an adolescent nephew who cannot be away from interactive media devices for very long without complaining of boredom. Mother intuitively suspects that Kara's fascination with a non-living interactive media device at the expense of human contact might not be good. So she limits the time at the computer and makes sure to have most of her special time with Kara away from the computer.

Comment: These precious experiences provide Kara with a plethora of essential ingredients and Growth Opportunities in just the right proportions and in just the right way. They are gigantic investments in all aspects of Kara's development, and they reward mom just as hugely. However, interaction with media itself is not essential for these experiences, but merely provides opportunities for mother and child to be together.

Mom is right to limit the time that she includes the computer in her special moments with her child. We do not yet know the long-term consequences of exposure to a non-living interactive media electronic device when not mediated by a loving adult. I suggest that early exposure in this manner can set the stage for the proper use of interactive media in later childhood and adulthood.

KEEPING IT "GREEN"

SEARCH WORDS: computer baby toddler DVD TV screen time

Baby's brain is indeed a precious natural resource, an environment of rapid growth and delicate balances worth protecting and keeping "green." Parents who understand this concept would want to present balanced Growth Opportunities to their babies to maintain their brain's ecology and avoid unnecessary risks and potential harm. Know your child. Her individual developmental needs should guide the content of interactive media. Some kids need and tolerate a lot of stimulation, while others become uncomfortable with just a little. Follow your child's needs. Be sure that media do not over-stimulate the child or demand responses that are too slow to keep her interest or too fast so that he becomes overwhelmed or frustrated.

Bottom line: know your own baby and watch her reactions. Do not overuse TV or baby videos, and avoid background TV, radio, or other media which significantly disrupt attention and the quality of children's play and other important developmental activities. When evaluating claims for a baby video product, remember that any benefits are limited by a baby's own brain equipment. Children's sensory and cognitive abilities develop gradually, and in the main probably cannot be substantially and enduringly accelerated. Do not let claims and testimonials get your hopes up that products claiming to be "educational" or enhance "learning" will actually provide any value or have any substantial long-term benefit, especially under twenty-four months.

The lines among Growth Opportunities get sharper with age. For simplicity, classify screen exposure according to its main purpose, although it may serve other aims as well. Know your child's needs, strengths, and challenges. Anticipate her next developmental steps realistically and base your expectations on what she can actually do. Know that her development unfolds in a manner generally similar to other youngsters, but its specifics are unique to your child. Reward always works better than punishment, approval more than disapproval, affection more than irritation. Reward effort as much as accomplishment. Avoid excessive praise for competencies already accomplished.

Family Relationships dominate during this period, taking priority and providing a framework, scaffolding, and context for all other Growth Opportunities. Nature is clever in her design—a newborn's smile is merely a reflex designed to seduce parents and recruit them to take care of him with confidence. The infant is basically only responding to the shape and features of any face, or would even smile at a simple drawing. Only by about three months does a baby actually distinguish mother and other close family members and react to mother's emotional state.

Kids as young as six months can learn to recognize words if taught lovingly by a parent. At about 7½ months, baby learns from you to recognize basic language sounds much better if she sees your face and background noise is quiet. Nine-month-olds learn the sounds typical to their language only during play interactions with loved ones, not media. Save energy and turn off these media.

Twelve-month-olds can show distress when another person is upset. They are learning to read and react to feelings. At twenty-four months she can start learning empathy and can now be introduced to the concept of kindness in pretend play.

Although intimate face-to-face contact is essential at any age, it is crucial for babies and very young children.

You can help your baby master her anxieties, mostly about losing you, with games like peek-a-boo or, later, hide-and-seek. Finding each other is a time for an exciting and love-affirming reunion. With time, she practices leaving you in wider and wider circles to explore other rooms in the house or even wanders off in a store, but is always counting on returning to you.

Young children are easily frightened, even from things we would not usually anticipate. At twelve months she can be visibly affected by what she sees on the screen. Get in the habit of choosing only positive, friendly, warm content, without any suggestion of separation and loss and violence, or anxiety or fear in any of the characters.

Baby begins to tell the difference between an object and its picture a little before twenty-four months. She feels these images as vivid, immediate, and real, because she cannot fully appreciate that some things are real and others only images or illusions until about four years old. Avoid violent news programs, cartoons, and loud sounds.

Although she does have night terrors; dreams and nightmares that adults experience probably begin around three years old, but do not get fully established until about eight years. She feels these terrors as real and cannot be calmed merely by words, even though you try to assure her these are not real. But your loving touch and soothing emotional tone really make the difference.

As her brain matures and mind develops, she continues to need a safe, nurturing, and sensitive family. First and most importantly, keep her safe and loved, so that she can trust you and learn how to know what trust and love are. Avoid frightening your child. Remember—the main purpose of interactive media sessions for the very young child is to promote development through an enjoyable interaction with the parent. She clings appropriately, thrives on your touch, and later desperately needs your approval and love. Of course, she needs your protection and supervision.

Socialization begins with family members, but widens later in the second year. She begins by interacting more actively with siblings and other family members, engages strangers, and can stay more easily with a baby-sitter. Within this cradle of relationships to others, her personality blossoms. Saying "no" is her affirmation of her individuality and a necessary step toward separateness and later awareness of the separate existence of others. She insists on initiating contact with others and directs her attention with greater intentionality. You want her to respond positively to adults like teachers and relatives, but to come to you or the proper adult when she feels fear, pain, and/or hurt. You want her to wander off less and learn to be more careful with strangers without mistrusting and fearing people unnecessarily.

Values Education begins during this period. In the first eighteen months, learning to love and trust and to see the world as a fundamentally safe and good place serves as the basis for later values. As her mobility increases her potential for getting into trouble, morality begins to be tied to parental approval. Wrong is whatever displeases you, and right is whatever assures her of your love. At about eighteen months, she looks to you for approval of her actions and learns how to see herself as good, competent, and proud. She also starts to know shame or guilt. She reads your tone of voice and other non-verbal behaviors better than you think, and she easily feels your disapproval with shame. To herself, she is either good or bad, lovable or worthless. Things are mostly black and white with little in between for her. She needs guidance in a firm but respectful manner. Appropriately discipline her quickly and firmly to teach her self-control. Focus on her actions rather than her as a person. You can expect her to have rough and solid ideas of simple rights and wrongs and to let you define these for her in her third year.

Becoming a consumer is an important part of Socialization in our society, and marketers deliberately and carefully try to influence children as young as six months. Such children can begin recognizing characters licensed from Entertainment and learning media. Marketers know that brand recognition is an early step to brand choice, preference, and loyalty. Marketers want younger and younger kids to reach for their product as Mom pushes the cart down the grocery store aisle. They capture such youngsters' attention with novel, quickly changing images that have color and movement. However, flashy graphics and rapid scene cuts can actually harm children's learning to pay attention by not allowing them to complete the necessary steps in the process. Their easily captured attention, development of strong preferences and stubbornness, and relative inability to tell fantasy from reality are all exploited by marketers who target this age group as consumers for their products.

Young kids are much more attracted to the media character associated with a brand than to the product itself. They typically like round, googly-eyed, smooth, gentle creatures with clean lines and interesting, but clear voices. Marketers work very hard to expose young children to images of these characters from early infancy so that the child will recognize it and have you buy the brand. Disney is a brilliant master of this phenomenon—who doesn't recognize and like Mickey Mouse and his friends?

Education Enrichment. Within the comfort and safety of your relationship, a baby seeks Education Enrichment eagerly and spontaneously. Now is the time for media baby food—easy to chew and digest and gentle on the body and mind. Her main education is to become a loving person. Education Enrichment is inextricably tied to activities with family and later with others.

Baby is wired to soak up new experiences and information and is superb at it. Her ability to recognize the difference between a thing and its symbol develops slowly. She tends to focus her attention on one dramatic feature of an item. It is also hard for her to delay gratification, so she wants what she wants when she wants it. Until twenty-four months, she lives in the here-and-now and has little concept of waiting. These qualities are often exploited by marketers and can lead to conflict with Mom at the grocery store.

Using a DVR, fast forward and edit out commercials. Always tone down volume, brightness, color, and contrast. Preview DVDs and save for later use what she will watch on TV through the use of DVRs.

No matter what her temperament, your child seeks mastery over intellectual and physical challenges spontaneously. Everything within touch, visual, or auditory reach can be fair game for learning, so stay alert for opportunities and satisfy the child's curiosity and need for repetition. For the most part, you merely need to expose him to opportunities in the course of daily life, encourage and then follow her lead.

Cognitive and sensorimotor learning is a natural and urgent need for babies. Soon after birth, the infant begins to have periods of alert inactivity when she is bright-eyed and peaceful, making sounds, moving her limbs, and looking around with an intent gaze. Baby seems truly present in these precious moments. She is studying her surroundings and might also be learning to be curious and to choose what she wants to focus on. She definitely does not need an intrusive and impersonal baby video during this important early state. I recommend that you let her have some of this time to herself, and otherwise enjoy her in gentle and playful interactions. Well-intentioned parents may think it necessary to offer her learning activities with TV and videos. Don't. There is little evidence that these will add anything more to her development than simple mobiles. But seeing your face does. Remember that throughout childhood, and even beyond, learning is enhanced when it is a social experience.

Learning how to learn can start at this stage. Gently train the child to focus for gradually longer periods and begin to play memory games and tasks requiring groupings and categories (for example, animals vs. plants). She should show pleasure from mastery and increased self-esteem. Cheer her gently when she stretches herself, persists, and then masters a new task, even if it is just to sit still a little longer. You want to encourage her ability to persist in the face of frustration, so model the solution or make a difficult task a bit more novel to reengage her and hold her interest. Remember that she learns by repetition and imitation, and have patience. Provide your child with opportunities to expand her skills, and bring pleasure from mastery, stimulation of imagination, and creativity. Recognize and encourage her originality and innovation. Respect her developing preferences—even stubbornness—especially when she directs her attention at will, a uniquely human ability that allows her to invent thoughts and actions, not remaining merely reactive to her environment.

Much of language development occurs naturally in the course of ordinary daily life in homes where family members talk to each other. No media can replace that. In the second year, especially as she begins to master walking, the world becomes her oyster. She explores it with intense curiosity and energy. She expresses her needs in words, not just gestures or nonsense syllables. She is progressing in her understanding of simple words and reacts to tone of voice and verbal expressions. She is also developing the ability to manipulate symbols in play and can join speech to make language. Her language becomes a powerful organizer of her thinking, behavior, and relationships.

Screen time refers to having a TV or other media in the foreground or background, and babies are indeed influenced by what they see on a screen. I will cut to the chase. Parents are busy people, balancing many spinning plates simultaneously. That is a reality. However, professionals often bow too easily to parental claims that they just don't have the time to tend to essentials in babies' care and must use TV. Well, I don't buy it. Don't have a baby if you cannot put in the time to make her your first priority and are not ready to sacrifice a whole lot of your convenience for her. Having said that, I know we do not live in a perfect world. What's more, babies are resilient and can grow well with good mothering that need not be perfect. In fact, some optimal frustration is healthy and leads to better development. But it must be optimal for the baby, not for the parent.

So what does a mother do when she needs some of her own down time, if only to take a shower? Leaving baby in her crib asleep after a meal or awake in a state of alert inactivity are the best choices. If baby is fussy and needs more stimulation and is not satisfied with the toys or mobiles in her bed, you may turn on the TV. However, make sure that the sound volume is almost on mute and that the video has no quick cuts and high color contrasts. Paradoxically, I have suggested to some parents that a shopping channel may be most ideal because it is peaceful, not dramatic, and slow in switching scenes, yet provides enough novelty and visual interest. (By contrast, watching a fast-paced music video, with quick cuts and a torrent of images, even at normal volume, is the worst option—for older kids, too.)

Learning word meaning vs. recognition can begin as early as fifteen months. At this age, she can learn from TV if an adult is facing her and speaking slowly, especially with Mom joining and helping. Beginning at about eighteen months, some babies can make sense of a rudimentary logical progression or story line, but the ability to understand a story that might interest adults does not begin until four or five years. At twenty-four months, children can learn from TV best if they are allowed some interaction.

Stay a little ahead of her and encourage your baby to stretch. Use words and simple sentences. Do a lot of talking with her and explain the world in ways she can understand and start echoing.

Here is some of what we know in general: At six months, Baby can see clearly and show sustained attention to a screen. She is attracted to a screen's ever-changing novelty, but what she actually sees or hears or what sense she makes of it—certainly her readiness to learn—depends on her brain's maturity and her social setting. Do not assume she sees or thinks what you see or think! At six months, she does not hear subtle differences in sounds and cannot easily localize them. However, at that age she can begin to imitate simple tasks she sees on a screen, and by fourteen months she can do so consistently, but needs more repetitions. Little is as yet known about how these abilities translate into actual learning from interactive media.

Studies have shown that more TV viewing is associated with less vocabulary and expressive language development. Other studies have shown that children read less with more TV viewing. Still others have found no such results. No similar studies have been done yet for interactive media, but we know that the quality of content and parental involvement can lessen damaging effects.

Attention is a complex brain function that makes learning possible, but it is itself learned in early childhood. Learning to pay attention is a difficult and draining job for your child. While locking attention on to something is the first step, there are more steps that follow. The toddler starts by looking, but can need up to twenty seconds to exclude distractions and find enough clarity in what she is actually seeing to focus her attention on it and to start having thoughts about it. Once the child "tunes in" and thinks, she generally can pay attention for twice her age in minutes. For example, a two-year-old's attention span is about four minutes. We know that attention problems follow traumatic events in early childhood. Doctor Dimitri Christakis and his colleagues at the University of Washington argue that TV watching at ages one and three interferes with learning how to pay attention so much that it leads to attention problems at age seven.

How you direct the child's attention is important for younger children. Looking at a picture book on your lap together, your child has an opportunity to learn and increase her attention span and impulse control better than anything offered by most screen media. Watch her face and wait. She can become quite absorbed. Let your child take her time studying what she sees. Let her set the pace and turn the page or change the screen display herself, and reward him for persisting with attention, with hugs and praising words. Name for her what she sees just before she is ready to go on.

Her sensorimotor skills are rapidly blossoming as she delights in movement and sensation. Through her infancy, she explores constantly, and can talk incessantly, with her gross and fine motor and eye-hand coordination and skills leaping forward. In her second year, input media devices should accommodate the child's fine motor and eye-hand coordination. For example, a touch screen is good initially. Consider reasonable, simple games that simulate and build on games already initiated by the baby, but with Mother being fully present for at least 80 percent of the time. For instance, Baby discovers that touching a screen or other device in a certain area, and with sufficient pressure and duration, elicits a response. Research would be required to show the best use of such a device and the long term benefit.

Let the child's attention span and interest guide you. Older children require more careful selection of content to maintain their interest. Avoid the pitfalls of being too pushy and ambitious, or of not being ambitious enough. Your role is to protect and guide the child and give lots of hugs and encouragement along the way. Do not leave the very young child alone, and keep the hardware out of reach when not in use. Look for her frustration tolerance and attention to increase, but she is limited by her immature brain. As she matures, her moods will become more even and less disruptive during longer periods of learning activities.

   Your own enjoyment of tending to her bodily needs—feeding, holding, washing, and changing diapers are essential early experiences for her respecting and enjoying her own body. Some like rougher play more than others. Some babies are more sensitive than others to touch, noise, or visual stimulation and can be easily distracted. It is best not to put too much on their plate. Keep radios and TV off when tending to them. "This Little Piggy" type games allow for joyful loving discoveries of her body.

Chapter 5: Learning What's Real (Two to Five)

The child continues to consolidate and build on earlier gains as he develops new abilities. The normal child squeezes Growth Opportunities out of every experience, according to nature's grand design for him. His more specialized brain feeds voraciously on the world around him as his curiosity drives him to find every Growth Opportunity in daily life.

He thus transforms himself from a baby into more of a child. To our delight, he becomes more like us as he learns skills and develops specific abilities. Growth Opportunities become more distinct: Family Relationships continue as the necessary scaffolding for Values Education, Socialization, and Education Enrichment, which are mostly entertaining and fun.

Except for the child with special needs or an unusual talent or gift (like the actual Mozart), or the too many children who are significantly deprived because of neglect or poverty, daily life provides sufficient opportunities for your child to find his own Growth Opportunities. There is little evidence that he will benefit significantly from interactive media.

Family Relationships continue to dominate, but your child accepts and seeks others at places like Day Care. He now has a clearer understanding of his place in the family and understands relationships. Your main job is to keep him safe and loved, so that he learns trust and love. Avoid frightening your child. Avoid media exposure to violence, including news and cartoons.

Teach your child by example that interactive media are family appliances. Begin to bring the whole family into the circle around the computer and enjoy slide shows, vacation planning, visiting with relatives, and other online and interactive media experiences together. For example, listen to audio books as a family. Work with your child to expand his world and become curious and aware of other peoples, religions, economic classes, and political conditions.

Around three years, his night terrors and daytime fears might increase; he will need your loving touch and comforting emotional tone as much as ever. Both day and night fears become more common, including those of robbers and kidnappers. Let your child continue to get comfort from his transitional object—a blanket or stuffed animal, for example—and his new imaginary friends. You want him to respond positively to adults like teachers and relatives, but to come to you when feeling fear, pain, or hurt.

Socialization opportunities and challenges become more complex at home and as the young child moves from play groups to nursery school and kindergarten. You want him to stay friendly and positive to others, while starting to form preferences in playmates. He progresses toward interacting cooperatively, but still puts himself and his intentions in the center of the universe. This is not yet selfishness or a character defect. He is on his way to playing with others, at least in a parallel fashion. Your child gradually becomes aware that he is a person, only as he becomes aware of his similarities to and differences from others and becomes curious about how people differ from one another.

He is open to *Values Education*, which at this point closely ties to Socialization and Family Relationships. He can learn patience, persistence, and to follow rules. TV shows that demonstrate values are particularly helpful. Watch these shows with your child and reinforce these lessons in daily life. Four- to seven-year-olds can learn leadership and problem-solving skills from TV shows like *Dragon Tales. The Wonder Pets* is another program that helps with learning values.

Education Enrichment is fueled by a curiosity that is partially inborn and partially developed by parents who have encouraged exploration, inventiveness, and originality in problem solving. By thirty months, children begin to show eagerness to read and ask repeatedly for the same book or show.

The mind evolves as the brain matures. Provide your child with opportunities to expand his skills and bring both of you enjoyment from his mastery, stimulation of imagination, and creativity. Recognize and encourage his originality.

Growing language skills enable him to define himself better and feel more and more his own person. You know he is learning, because he comes up with original names, descriptions, and colors. Your child is also beginning to represent mental images by drawing, drawing circles and lines, then progressing to rough human faces and protruding limbs. His world is larger because he is beginning to form an inner map of reality that enables clearer thinking as well as the storage and retrieval of memories at will.

Encourage singing, using rhythm, naming colors, and looking at illustrated stories and poems. You want him to enjoy looking and describing what he sees and what he finds beautiful. Select animal pictures, simple human faces, or cartoon figures. Then progress slowly to more language-based and multimodal output. Be sensitive to what captures the child's attention as a marker for what works. Encourage keyboarding using a keyboard with specially marked keys to create sounds, words, and simple melodies.

As your child gets older, your choice of media content becomes increasingly important. Much content is available online. You will know you are successful when your child shows curiosity and pleasure, and asks for more similar activities and interactions. Children can come up with novel kinds of interactivity. Let them. Use only interactive media that have carefully-tested curricula beginning where your child can master materials that move him through a progression of experiences to expand his abilities, much as TV's *Sesame Street* and *Blue's Clues*. His ability to understand what the future brings is increasing from minutes to hours, so let him know ahead of time when you plan to have a media session.

Children older than twenty-four months learn best from shows that have less frequent cuts between scenes, a linear story line they can easily follow, slow speech and exact correspondence between words and what is on the screen, and slow progression with much repetition. Look for media products that are carefully crafted and tested, like Sesame Workshop products. Of all TV and video products, *Blue's Clues, Barney, Dora the Explorer, and Dragon Tales* have been proven to have educational value and can increase vocabulary. For older children, convey your expectation that such shows can be for education and not just for fun. Be present and pay attention. Participate in learning together with your child.

His motor skills are increasing as he regulates his balance and movement more gracefully. You want media that will help him enjoy and control large and small movements, dexterity, and eye-hand coordination. Let the child experiment with the mouse, joystick, game controller, and even keyboard. As he gains skills, give more challenging tasks, progressing to the use of painting or graphics programs. However, do not neglect manual drawing and painting and the muscle training they provide. Dance pads and other human interface devices can be fun and beneficial. Do not use easily breakable hardware.

Teach him how to make good choices of media by offering him what is good for him and rewarding good selections. You also want your child to begin building on his independence in eating and bathroom and hygiene habits. He is proudly succeeding in controlling basic bathroom functions and eventually achieves full control. Turn off all media when she is learning these functions and be fully present.

Regardless of gender, you probably want your preschooler to become comfortable and familiar with the technology that will later pervade his life. You will want him to make technology his servant, not a master; a tool to enhance relationships, not to impair them.

As long as you keep these principles in mind, you can start exposing your child to computers and online and interactive media experiences. Children develop preferences for characters they see in the media and form brand loyalty to the products they adorn or endorse, so make rules like "We don't buy what's on TV just because it is on TV."

Entertainment becomes more of an end in itself during this age group, but it also promotes other Growth Opportunities. Older kids interact with media in more complex ways and are beginning to play video games.

Video games can increase spatial performance, image-reading skills, and visual attention. Some of these skills are measured in IQ testing, and it is possible that some IQ scores increase. Academic performance can be increased with proper use of good software. Children in this age group are more vulnerable to violent action games: Watch out for these, and prevent your child from playing them, enforcing a rule that you do not permit them. Emphasize entertainment where people are treated with dignity and kindness.

RECOMMENDATIONS

Although interactive media has become a powerful force in the lives of most developed-world children and their families, studies on their impact on effects are surprisingly few. However, studies about the effects of television viewing are suggesting the correct approaches. The American Academy of Pediatrics has recommended no more than two hours daily of screen media. Twenty percent of children of this age group watch more than two and a half hours daily, and many have sleep, attention, aggression, and other behavioral difficulties.

I recommend no more than one hour per week total media time initially, with no longer than twenty minutes per session, increasing to a maximum of three hours a week, with no longer than forty-five minutes per session by age four. I suggest that a parent remain nearby and alert ten percent of the time and stay directly involved, preferably face-to-face, over ninety percent of the time by age five. As much as possible, pay attention to what your child gets in Day Care, since most caregivers there may not be sufficiently informed about your child's needs. Try to choose a facility accredited by the National Association for Education of Young Children (NAEYC).

For further details and instructions, please refer to the recommendations, table and charts given in Chapter 3 on pages 36 – 37 and see www.mydigitalfamily.org.

EXAMPLE

Ben is a loving and well-developed child. Beginning at about age two, his parents started to follow these guidelines to prepare Ben for a world full of interactions with media. They were aware of the importance of this period in the growth of Ben's brain and the development of basic personality traits and cognitive abilities. They wanted to make sure they did everything to prepare him for life in a technology-rich world full of great opportunities, but also full of violence, consumerism, and exploitation. They were aware of the power of all media, especially interactive media, to influence the development of children, and wanted Ben to receive as much of his Growth Opportunities through them at home. They wanted him to start learning good, balanced habits for using interactive media for its benefits and avoiding its drawbacks.

Mom began Ben off with a variety of interactive media, including handheld devices, online sites, and interactive software. Mom spoke and sang together with Ben. Dad soon joined the fun. It was a chance for the family to learn while being together.

Mom and Dad found the right pace for Ben. They were alert to Ben's preferences for looking and listening and saw firsthand how stubborn he could be when he liked a video display. Mom and Dad were sensitive to Ben's limited, but growing ability to focus his attention as they patiently waited for him to make sense of what he was seeing and hearing, hoping to maximize his ability to pay attention. They watched his face and eyes to detect how he was doing when viewing a screen.

From the beginning, even before Ben was born, his parents decided how to manage interactive media in their home. They made the home computer into a family appliance. As a baby, they focused on face-to-face playing and personal interaction. They shielded Ben from seductive, colorful, fast-moving media content. Ben did not have any screen time alone and no background TV. At other times, the audio was turned down and video color and brightness muted. They were quite aware that such displays are intended to capture his attention and to get him to influence Mom's buying decisions. They worried that too much exposure could hurt their developing son's ability to learn to pay attention.

His parents also saw teaching Ben to be a discerning consumer as their responsibility, and they started as early as the marketers. Saying "no" in the grocery store is the best way. Mom even removed foods from packaging before serving them so that Ben would not associate the packaging, commercials, and nutrition. Later, Mom and Dad repeated, "We do not buy this just because it is on TV or the computer."

His parents slowly taught Ben in a fun way that media content like commercials and ads are pretend and even fun to see, but they are not for copying, emulating, or obeying: "Do we do this for real?," "Do we buy this because it is on TV?"; "Do we listen to the man on TV?," "Do we like this cereal just because Freddy the Frog likes it?" They tried to shield Ben from violent cartoons, news, and other media content, but inevitably, Ben's parents needed to mediate his understanding of the difference between fantasy and reality.

By age four, Ben's parents were well on their way to teaching him the differences between reality and fantasy. It was fun taking turns asking each other, "Is this real?," " Is that pretend?," and "What do we call this—a phone or a pretend phone for Ben to play with? A nice picture of a dog to look at or a real dog to pet or that to pet? A toy dog to cuddle with? Is it alive or fake?" His parents were careful to stay within Ben's ability to understand and enjoy the game. They were also careful to encourage Ben to enjoy playing with pretend toys and even to become so engrossed he lost himself in play.

They wanted to teach him their values, including the need for kindness and consideration and their disapproval of violence, even in cartoons: "Does the mouse have an owee?," "Does the owee hurt?," "Can the mouse cry?," "Is the cat good today?," "What else can the cat do instead of scaring and biting?" Ben's parents approved of non-exploitative displays of kindness, forgiveness, consideration and peaceful problem solving: "The cat is nice for sharing his cheese," "Frody frog is nice to let the fly go," "How can Frody be nicer to all the other frogs?"

Comment: Ben's parents were diligent in using media to maximize his development. He developed better visual cognition, eye-hand skills, computer literacy, and the ability to process visually presented text and other data. His parents wondered whether games may provide Ben with opportunities to develop brand-new combinations of perceptual motor skill sets that organize new mental schemata and brain circuits as he matured. They tried to teach him the importance of patience. As he developed increasing abilities to understand, they were careful to teach that fantasy and play can be enjoyed, encouraging fantasy play.

His parents kept control in their home over Ben's development by using media to maximize Growth Opportunities and teach him good habits of mind. They introduced him to the pleasure of collaborative activities.

In doing this, they also shielded him from the unwanted influences of strangers in the media. For the sake of Values Education, they also took every opportunity to teach Ben to avoid the influence of the media in modeling behavior, including violence and excessive consumerism.

So caught up in the intense media barrage, we hardly notice it. But small children can be overwhelmed and their development impaired unless protected. Growth Opportunities for very young children and their families can be substantial when mediated and interpreted by a loving family member and can be extremely valuable as preparation for life in a technology-rich world, even beginning at a young age.

Sitting and enjoying is great for Family Relationships and a way of conveying the importance of family and the value of interacting together.

Chapter 6: Collaborating with Your Kids (Five to Eight)

Baby is now a little girl, old enough to leave home for a large part of her day and spend it productively in school. She is making gains on top of earlier developmental accomplishments and is very different than who she had been a year ago.

Because interactive media are increasingly important to children in this age group, putting in place a Media Plan becomes more urgent. Introducing older children who regularly spend time online, use mobile devices, or play video games to the Media Plan is more difficult than with younger children who have not yet established their own habits. You might very well run into protests.

Explain the concept of nourishment and a balanced, healthy nutrition plan. Explain that it is your responsibility as a parent to look after her and that you do not intend to merely control her or question her judgment. Negotiate whenever you can, rather than overpowering the child. Approach the matter gently, increasing firmness gradually. Be prepared for only partially successful results with many older children. However, do not lose heart. Merely conveying your values and preferences clearly and repeatedly can often sow seeds that will influence your child's eventual choices.

You can ask the older child for her preferences, but make the final dietary balance and menu decisions yourself. Often, you can expect that the child can join you in food preparation and cleanup. You want the child to learn through an apprenticeship of collaboration and then slowly take over this function, especially in preparation for more independence in adolescence. The same holds for the Media Plan and menu planning, and there is every reason to expect that you will get at least some positive results. Include in your plan all forms of interactive media. It will be much more challenging to impose a Media Plan on your older child, had she had not been exposed earlier.

RECOMMENDATIONS

SEARCH WORDS: computer Internet kids

Family Relationships continue to hold a central place in any approach to interactive media. Your child now has distinct relationships with family members. She admires and emulates the same sex parent as a hero and model, As she solidifies her gender identity. Much of the core of her personality is already established. She is interested in caretaking, dependency, or emotionality, as are most girls her age, and not in physicality, or roughhousing, and conventional culture encourages these traits. She is very conscious of your approval and feels shame and guilt easily.

Your role becomes mentor and teacher, in addition to nurturer. Fathers and sons and mothers and daughters naturally seek each other out, so spend as much one-to-one time with your child as possible, including interactive media time.

The older the child, the farther away the parent can be from the computer. The child can use the computer as a word processor or calculator alone. I suggest strongly that you don't leave the child alone to roam the Internet freely or allow Internet access from a mobile device. Invite other family members to join your sessions with their own input devices.

Your child is getting better at negotiating, but still sees things as black and white. I recommend that parents take care to follow through, providing rewards and incentives and enforcing rules and giving punishments. Be consistent but avoid arbitrariness. Explain how you have arrived at the punishment. Discuss his strengths and challenges realistically and without inappropriately boosting and certainly avoiding harming his self-esteem. More and more, allow the child independence to learn from mistakes, accept responsibility, apologize genuinely, and repair damage he may have caused. Avoid buying him out of trouble or allowing him to assume the role of a helpless victim.

Expect that he will sometimes invite you to join him or help figure out a problem with the interactive media equipment, and will even reveal his excitements and discoveries. Your child still accepts your guidance easily, but can later become more assertive and stubborn. I recommend that you respond to your child's growing need for reciprocity and mutuality in his collaboration with you. Privacy is his way of saying he is his own person, and it becomes a right he demands and oftentimes deserves.

Socialization becomes more important to your child, and he has plenty of online opportunities. He is now capable of functioning among others. Increasingly able to take initiative and assert himself more effectively, he is learning to play and work both cooperatively and independently and feels good about the results. His solitary play is still important; it is becoming more complex and includes reading or drawing. Play with others is cooperative and mutual, no longer parallel in nature, and eventually the child feels like a member of a group. He is more realistically confident and optimistic and is starting to strive toward ideals based on expectations of himself and others.

As a child progresses through this phase, she increasingly values important and distinctive emotional and task-oriented ties with people outside the immediate family, such as friends, teachers, coaches, and tutors. As her own idea of herself changes, she is also becoming more complex and refined, able to show both sympathy and empathy toward others.

Her awareness of the feelings and needs of others is increasingly based on a sense of mutuality, a trait better developed in girls. She knows that people can feel more than one emotion at the same time. She is loyal to her friends and increasingly cares more about the judgment of her peers. She wants to belong, and can be easily hurt by exclusion, as her peers form cliques. She expresses her emerging self through joining her own choice of clubs with friends. Later, her desire to belong to cliques is invested with intense emotional energy and is based on social status. At this age, children can be cruel by excluding others, a worthy topic for Values Education and application of the Golden Rule.

Like other forms of play, anonymous virtual reality can be a laboratory for practice and experimentation in various roles and identities. Disembodied text-defined selves, online names, and role-playing avatars (figures that the user constructs to appear on the screen for him) interact with similar others to experiment with friendships, bullying, and group identity in interactive virtual communities. Some online communities can bring out better behaviors in their members.

Boys become more boyish and girls more girlish in many ways, and they make strong friendships with members of their own gender. At this phase, girls begin to seek fashion, entertainment, social networking, and grooming sites. Boys like action and fantasy games that can involve peers around the world. Graphic violence with pointless wanton mayhem is not good for children. Monitor video games and forbid playing with strangers online. However, keep in mind that kids have played Cowboys and Indians and other adventure games for ages to work out anger and mold untamed childhood aggression into mature qualities of healthy competitiveness.

Such games, digital or face-to-face, played in moderation, are necessary for healthy development. I suggest that you limit these to less than twenty percent of interactive media time. Visiting other homes as a family, agree ahead of time what kids will do together and have them bring media gadgets that invite social interaction. Do not allow personal video game consoles or smartphones that isolate and prevent social interactions.

Games and social networking are becoming more and more important in your child's relating with her peer group. Learning cooperation and collaboration with others interacting with media is fun and provides all other Growth Opportunities. Allow some sites once you have explored them yourself, but monitor use closely and limit the time to no more than twenty percent of the time. Interactions with other persons can extend to occasional use of e-mail and instant messaging.

Although most older kids use interactive media for communication and relationships, studies have shown decreases in social and psychological well-being, especially in the first two years of Internet use. Virtual reality can be confusing but quite powerful to children who cannot distinguish it from everyday reality. Violent computer games can spawn violent behavior in some children.

Kids are inventing new ways of integrating interactive media into their lives, often leapfrogging industry. They invented texting and the use of Net lingo abbreviations, and the practice is working its way into cell phones. Younger kids are picking up on these practices both online and via mobile phones. Social network sites like facebook.com and myspace.com are becoming more popular among young children. Superficial banter and small talk texting are common, as children as young as six learn the nuances of language and human social contact and are writing to each other more than ever.

Abbreviations keep the right rhythm and speed better than full words and sentences. To maintain a quick back-and-forth, they break a longer sentence into shorter fragments. They use abbreviations—lol (laugh out loud), gtg (go to go), boctaae (but of course there are always exceptions)—to substitute for non-verbal cues and to keep a natural conversational rhythm in texting. This is now a main reason for the online lexicon. Informal usage is entering our lexicon rapidly, already influencing the writing of English. I recommend that you limit the replacement of face-to-face interactions with texting. The art of letter-writing, written expression, and language development is not always best served in these types of online exchanges.

The practice is mostly innocuous. Major secrets are rare in this age group. But remember that a shy child may prefer text messaging to avoid more difficult face-to-face encounters. Online blogs and chat rooms are appearing much like the after school hangouts in older generations. Texting might be more comfortable for kids with disorders of Socialization like autism, Asperger's syndrome, depression, and shyness. Kids can perpetrate or become victimized by cyber bullying. If you consider how you limit freedom and privacy in their other social activities, you will realize that they require chaperoning online, too.

Values Education becomes a more important Growth Opportunity in this phase and requires continuing creativity and innovation from parents. Closely tied to Socialization, it continues to require extra efforts from parents. Use Socialization experiences as laboratories for your child's Values Education. Your child now values rules dutifully and literally simply because they are rules. He defers to authority, rigidly expecting others to do the same. Fairness, right, and wrong become powerful ideas that are personally important to him.

He still does what is right because it earns him love, but also because he feels good about it. He fears punishment and withdrawal of love for doing wrong. He can be encouraged to take responsibility for acts of kindness to others. He may include an idea of God in his thinking if this is a learned part of the family's values. He may extrapolate that God expects good behavior. Emphasize opportunities, social responsibility, and kindness in the more complex social interactions of his daily life and interactive media use.

Older children can be allowed to have more choices as they exercise good judgment. Start teaching safe practices to the older preschooler. Repeat the rules and post them. Begin to teach the child respect for the power of interactive media, both good and bad, and the need for judgment and caution online, just as in the street or in your kitchen. Showing compassion, empathy, kindness, and respect are important Socialization values to breed in younger children but are especially important now. Start to teach values in earnest with sites like the Red Cross and others that help people. Find interactive media opportunities that help the child develop values consistent with your own.

Children who appear to be doing homework using online resources may stray into other activities, such as instant messaging and games. If the honor system does not work, parents may have to install filtering and timing software to prevent unauthorized activities.

WE LOVE YOU, _____

WE WANT YOU SMART AND SAFE
- Use your common sense.
- What you put online stays online—you can't ever take it back.
- Never give out your name, address, phone number, your school, or photos.
- Follow all the rules of the real world in cyberspace, like no swear words.
- Treat all people politely and kindly.
- Never bypass or weaken our computer protections and security.
- Never let others know your passwords except us.
- Never go to meet someone you met online.
- Never answer anything strange or weird.
- Never agree to be sent additional information.
- Do not believe everything you see or spend money online.
- Don't snack while online
- Get our permission before downloading or uploading.
- Only go to sites approved by our firewall or a safety rating service.
- Never send or post photos of yourself or your family.
- Let us know right away if there is something different or wrong when you are online.

Education Enrichment at home now supplements school activities. Your child's sensorimotor and cognitive growth originates in a brain that is now ninety percent the size and weight of a grownup's and is increasingly powerful.

Your child progresses from weakness to strength in logical thinking, but remains literal and concrete. As his ability to manipulate and transform symbols into thoughts begins to evolve, he becomes aware of himself as part of a larger world, rather than its center. His language skills allow him to speak smoothly and use many words. The blossoming of his imagination and creativity is helped by greater discipline and capacity for attention.

He loved imaginary play with toys earlier in this period, but may now shift more to reading. He has a greater eye for detail and enjoys collecting. Encourage imaginative play with toys that capture her time and attention. Encourage such play online if possible, but beware that most video games have rigid formats and require little else but quick reactivity, allowing for little imagination.

Stimulate your child's thinking as you work online together, and encourage independent, detailed analysis, and descriptions with questions. The more mental energy a child expends, the better he learns to think and to enjoy thinking. For example, have your child explain why he likes or does not like a site. Encourage him to describe what he sees. Nurture the child's creativity, imagination and mastery of difficult tasks. Encourage opportunities that elicit curiosity and excitement. Work with your child to expand her world.

Encourage alone time without activities, electronics, or socializing. Be realistic in giving feedback. Avoid linking performance too tightly to self-esteem. When you give negative feedback, be careful, especially with girls, to avoid demeaning them.

She can balance herself and move gracefully as her gross motor functions become finely tuned. Her handedness emerges for all to see as she holds a crayon or throws a ball. Her fine motor skills and eye-hand coordination allow her to tie her shoes, write and draw more effectively, and learn to ride a two-wheeler. Special intellectual and physical talents become a source of pride. Change the physical arrangement and input and output devices you use as the child is more independent and facile with hands and fingers. Your child can operate any of the input devices and use any output device.

She can increasingly take care of her own body—bathing, dressing, and tying shoes—and is beginning to appreciate privacy. She localizes pain better and understands the function of doctors. She is able to appreciate beauty and creates it in drawings, use of language, dancing, singing, and playing a musical instrument. As her special talents emerge, she tries to develop them with a parent's encouragement and her desire for approval. Increasingly productive and creative, she now cares more deeply about schoolwork and other disciplined activities. Your child takes more initiative and is assertive, industrious, productive, and hopeful of success.

Stay involved, informed, and concerned, but allow your child increasing independence and responsibility for school life. She must learn for herself the consequences of slacking off. Do not rush to save her, but rather communicate that she is a complicated person in her own right and needs to learn responsibility.

Education Enrichment can now also be guided at school by teachers and librarians or by the discoveries the child makes herself. Enrichment can be an independent activity for many children, or an opportunity to collaborate.

Of course, you can work with your child to improve academic performance. For example, your child may link to the teacher's recommended sites or use search engines to find credible information. Some teachers already use a site to post homework assignments and grades. Remediation is an obvious way parents can help. Children's motivation to learn can increase with attractive, dynamic graphics and enjoyable educational video games.

It is at home that children usually learn their most enduring habits. Go beyond your child's school curriculum. His fine and gross motor potentials are becoming very finely tuned, and he may take seriously a sport, performing art, or other talent development.

Ask his teacher or librarian, or even a special tutor, to coach you in guiding your child's online exploration of broader areas to complement the curriculum. His schoolwork will improve. Researchers at Michigan State University have found that home Internet use appears to increase reading performance, especially in those lagging because of poverty and poor opportunities.

Welcome opportunities that stretch your child's development. Learning how to learn in the manner appropriate to this age group would include building on prior challenges, and adding mutuality and a more sophisticated understanding and analyses of the media's content. Pleasure from mastery and self-esteem should come not only from skill in games, but also from learning and mutual enjoyment.

For example, the range of your collaborative activities can expand. Encourage volunteerism by bringing your child along when volunteering as a Big Brother or Big Sister to a less fortunate child. Start developing your child's interests in the rest of the world.

Few children require much help in finding entertainment. Use Entertainment as a reward. Be explicit in rewarding the child who shows good judgment with extra Entertainment time. Your daughter now seeks entertainers to define her own culture and to specifically distinguish her from her parents' generation. Join her online or in listening to music and become part of this experience.

Older children can enjoy massively multiplayer online games. These MMOGs provide opportunities to interact with children in other countries, promoting respect for diversity. Left to themselves, older kids have always preferred to snack on Entertainment. Mental health experts agree that the harm of violent video games, especially to the more vulnerable child, outweighs by far whatever small advantage they may provide in fine motor and eye-hand coordination. Like bullying, violent games must be banned from your home. If you hope to raise good kids, I believe that your only tenable position is to speak against these games and seek alternative Entertainment for your children. As a doctor and expert in children's well-being, I cannot imagine any good outcome from such games.

Media literacy (SEARCH WORDS: media advertising target baby "market segment") can be taught in greater depth to this age group. Marketers have targeted this group since the 1980s, realizing that children's attention can be captured and trained increasingly by interesting and challenging content.

Both at home or while visiting friends, children can enter into commercial and other relationships that expose them and the family to danger and identity theft and fraud. The 1998 Children's Online Privacy Protection Act enacted by Congress empowers the Federal Trade Commission to forbid online data collection and the overt targeting of kids younger than thirteen. However, they are still aggressively targeted by marketers.

As kids progress through grade school, they increasingly crave social status and acceptance by peers through dress, jewelry, and ownership of the latest and best. They have definite tastes, and vast purchasing power, besides influencing family purchasing decisions. Along with peers, they can be seduced into intense brand loyalties and are first users of new products flooding the media. Boys commonly crave digital devices, while girls go for clothing, makeup and accessories. Peer-to-peer marketing also captures consumers who, pyramid-like, spread the seller's message.

Continue refining your child's identifying, understanding, and responding to media marketing. At this age, a rule such as "we don't buy things from TV or online just because they are there" may still be effective. Help the child understand the difference between wanting and needing. Video game makers and retailers are striking deals to make branded merchandise an essential part of the players' experience.

PARENTING BASICS

BULLIES AND THE BULLIED

The bully is a predator who uses physical violence or verbal threats to frighten and terrorize. He dominates, degrades, and humiliates a victim to gain a sense of power, lessen his own fears and insecurities, and increase his own self-esteem without regard to the damage he inflicts on his victim. According to one survey, a majority of kids have been targets of online bullying and most are aware of this behavior by their friends. Basically, bullying via interactive media is neither uncommon nor much different from its other forms. Anonymity in cyberspace enables the cyberbully to hide, while his attacks can be no less painful.

Most children, especially boys, have been occasionally bullied but commonly cope successfully by avoiding or standing up to bullies. However, the child who is bullied often requires identification and urgent attention from caregivers.

Cyberbullying is a form of verbal bullying that can be just as damaging as verbal or physical violence. Often, it comes from a child's social network, but anonymously, adding the isolation caused by suspicion and mistrust of his friends to the burden of being bullied. It is a form of violence that can beget violence and must be taken seriously. It is not just "kids will be kids."

Bullying can begin in preschool-aged children. But as children grow older, bullying becomes more serious and is increasingly concealed. It often comes to the attention of parents or school authorities long after it has started, because children are reluctant to "snitch." Unfortunately, in some American subcultures, bullying is actually an established, well-developed part of the social fabric. Both bully and bullied often desperately crave positive attention at home, especially from the same gendered parent or other mentor. Both may benefit from after-school or special school programming.

A bully is a predator who should be stopped immediately and monitored closely. Clear and consistently enforced limits and appropriate consequences should be quickly applied. Once assessed, the bully and his family can receive appropriate psychological help. A bully may need to develop increased social awareness and sensitivity to others, acceptable ways to increase self-esteem and channel aggression, and an identity not based on aggression alone.

Ideally, the bully could be brought to a point where he understands his obligation to apologize meaningfully to his victim and actually does so without coercion. Older bullies may be asked to perform community service. Bullying can be a traumatic experience for the bully who experiences remorse, shame, and guilt. Such bullies, who are also followers or act impulsively, are easier to treat and have a better prognosis. The bully who is a leader or deliberately cruel, or suffers little or no remorse, shame, or guilt, is much harder to rehabilitate and can be at very high risk for future criminality.

Assure the bullied child's safety. He may need help to develop self-esteem and to learn specific strategies to stand up for and protect himself. His other needs may include learning social skills or appropriate assertiveness or self-defense techniques and their deterrent value. Depression may require treatment by a child and adolescent psychiatrist.

Educating communities about alternatives to violence, tolerance for diversity, compassion for the weak, and promoting opportunities for healthy moral and social development are other aspects of a general approach to preventing and intervening with bullying. Children should be taught that they have a right not to be hurt or injured by others.

By reporting threats or harm done to another, children could learn to practice social responsibility. However, they should not be the sole sources of information or complaints about bullies, since it is the obligation of adults to protect children and prevent violence by identifying and monitoring bullies. Because it takes considerable judgment and maturity to weigh the social good of protecting others against the social bad of "snitching," how to decide when to tell and when not to tell could be regularly and thoroughly explored and discussed as children grow older.

It is the responsibility of parents and school personnel to provide children with an environment that does not tolerate bullying or any other form of violence. Although revenge by bullying victims has received attention as a cause of school shootings, let us not forget that bullying is itself a common form of violence. Parents should examine their attitudes toward bullying and regularly query children about their experiences at school and in the community. Schools and communities should assess the pervasiveness and causes and review policies and procedures regarding bullying.

VIRTUAL COMMUNITIES: FORWARD TO THE FUTURE

OURCLUBHOUSE.ORG

In addition to an online family site, a group of children in this age group can create their own special address in cyberspace with an adult webmaster who sponsors and monitors the password-protected site. A neighborhood, church, special-interest group, a group of friends can deck out its own gathering place. These can include a page for Media Plan resources and links to sites providing Growth Opportunities and sample weekly menus for this age group.

Kids will adopt procedures for accepting new members, developing rules and consequences for breaking them, and making changes. Links to their own page at their family site can be added, along with chat room, webcam, voice, or texting functions. Kids can post photos, news, announcements, poems, stories, essays, and music, movie, website reviews and other information.

Children from other neighborhoods, cities, or countries can be members, and clubs can find sister clubs in distant places. The club can undertake special projects to help other clubs or its members. A club can be a chapter of a larger national or international "mother club" that provides member clubs with advice, templates, connections to one another, and other resources.

A child can belong to more than one virtual community, much as he joins several clubs. For instance, he may join a group of collectors that buy and sell on eBay, or like-minded fledgling athletes or poets, kids who want to learn a new language, or children who belong to the same church. Parents may have a section to themselves for exchanging information and self-help.

As children grow and develop, they may change the form and content of the site to suit their evolving needs, or the club will naturally turn over to younger members as older members move on. This group of kids can stay together through adolescence, young adulthood and even beyond, enriching their lives with enduring friendships.

The possibilities are endless and will grow as Internet and other technologies grow.

Sociologist Dr. Juliet B. Schor has studied a slightly older age group. She found that the more materialistic and involved in consumer culture a kid aged ten to thirteen is, the more likely the child is to suffer from depression, anxiety, low self-esteem, and psychosomatic complaints.

Be patient. The urgent barrage coming from a child who craves the "latest and best" digital device is often the result of the intense pressure he himself is receiving from the media and his peer group. Be consistent and fair, but maintain your parental authority. Teaming with your partner can help you stand your ground. Remember—although the child is testing you mercilessly, he is also relying on your firmness and consistency for his security. Discuss how marketers segment markets and capture their targets. Encourage more sophisticated understanding and analyses of a site's content. Emphasize enduring values to counter the materialism and consumerism in your child's world.

Huge sums pour into chat rooms and social network sites by merchandisers. Such sites are often stealthily data mined as focus groups to determine preferences. Market researchers harvest data from the patterns of keyboard clicks. Marketers create attractive virtual malls where kids can visit stores and interact with others.

To help a child understand and minimize control by ever-present and powerful commercial forces, parents can encourage curiosity and independent critical thinking. They can teach an eight-year-old how advertisers intend to influence buying choices, the necessary place of advertising in a free economy, the sometimes fine line between ethical and deceptive practices, and the commonly-used methods of targeting demographic market segments by age, race, and gender. Parents can teach their older child to critique commercials to discover their intent, techniques, and targeted groups. Kids can also research this topic as part of their Education Enrichment Growth Opportunity activity.

LEARNING SAFETY: COLLABORATING WITH YOUR KIDS

Children learn best from tangible experiences. Here are two ways that you can teach (and learn) about safety and beneficial use of interactive media.

Fun for the whole family : A BOARD GAME TO TEACH AND LEARN SAFE AND PRODUCTIVE INTERACTIVE MEDIA USE

   a. Alter a game that is already a favorite to now include interactive media benefits and risks.

b. Start with a board game, such as Candy Land or Monopoly, that suits the age of your child. Choose a game that has reward squares that advance the lucky player and penalty squares that slow down the unlucky. Include cards drawn from a pile that also provide penalties, rewards, or protections. Include play money, too. You will use the board, playing pieces, draw pile cards, and dice or spinning wheel.

c. Have several pads of different colored small sticky note paper and colored pens on hand. Use small stickers to decorate and personalize the note paper.

d. Cut the sticky paper to fit the squares of the board game. Fit paper of different colors for hazard, reward, and neutral squares. Cover many squares of the board. Similarly, you may fit cut-outs of note paper on draw pile cards.

Although you will now transform the original game by changing the action, you will keep the original rewards and penalties:

a. Make a list of positive interactive media behaviors that can bring rewards, such as: effective uses of safety measures; following the Media Plan; avoiding giving private information online; or a benefit from Growth Opportunities. Arrange these in order of importance. Get more ideas from this book.

b. Repeat for undesirable interactive media behaviors, such as snacking at the computer and becoming obese, wasting paper and harming the environment, playing violent video games, and meeting strangers online.

c. Write the positive interactive media behaviors on sticky notes covering reward squares or draw cards of the original game. For example, write, "Your caution thwarted a spam attack" over the original square that reads, "Advance to GO and collect $200." Or, "Your up-to-date firewall blocks a spyware attack" over the square reading, "Advance twelve spaces;" Other positive interactive media behaviors can include spending an hour with grandfather visiting his hometown site or visiting the church site. Be creative and brainstorm. For example, in Monopoly, a hotel on Boardwalk stands for avoiding violent video games for a month, a house on Pennsylvania Avenue for updating firewall software, while a house on Mediterranean Avenue for avoiding snacking while online. Drawing cards like a firewall or an antivirus card can provide protection.

> d. Repeat for undesirable interactive media behaviors. Write on sticky notes covering the penalty squares or draw pile cards of the original game a negative event, such as causing identity theft by carelessly giving information online, becoming an unkind person from playing violent games, becoming addicted, becoming a bully, causing a computer crash from a virus picked up when software was deactivated, and causing environmental damage from wasting paper.
>
> Play according to the original rules by keeping the existing structure of penalties (moving backward, going to jail, and waiting until a number is thrown or a color spun) or rewards (moving forward, collecting money, sending your opponent to jail, etc.) of each square and draw pile card of the original game. For example, a player may blunt a virus attack when his opponent draws "Your opponent is attacked by a virus and loses five turns" with a card reading "You have good antivirus software."

Interactive media in your child's room? This is a question of family preferences and developmental readiness. In a way, parents have been asking the same question for decades about television sets. Research has shown little benefit and some harm to a child from TV watching in his bedroom. For instance, one study showed that kids who have a bedroom TV at age five and a half have sleep problems and are emotionally more flat. In recent years, bedrooms have become media centers that include other interactive media as well. Computer use for schoolwork has become necessary and commonplace as computer literacy is crucial to functioning in today's world. For these reasons, we can understand how an offline computer may be placed in a school-age child's room as early as six, as long as games and other Entertainment are supervised or blocked and Growth Opportunities maximized. But an online computer and other interactive media devices are totally different.

Online privacy is not generally appropriate until mid-adolescence, beginning about twelve to fourteen, no matter how insistent the child.

RECOMMENDATIONS

The Media Plan I recommend for an early school-age child is more complex and changes dramatically during this phase. By age eight, in addition to homework and other school-based computer time, a total of about five hours a week is best, with no more than one hour at any one sitting.

By age eight, the child may begin working with interactive media alone for a small amount of time if she is responsible and trustworthy. I she knows how to tell time, the child could take personal responsibility for adhering to the Media Plan. If not, take this opportunity to teach her to tell time, using the incentive of more independent online time. You may put a clock nearby, with or without an alarm, or on the screen (see the control panel date and time function). I recommend that a parent be present and alert nearby for over forty percent and directly involved fifty percent of the child's online time.

I recommend strongly that younger children in this age group continue to focus on Family Relationships, but the proportion of time spent on Socialization and Values Education now increases, while Education Enrichment remains the same. Remember, the child is already getting online educational activities in school and Entertainment while visiting friends. The increase in total time will result in increases in all categories, although proportions will remain the same.

The total times I recommend for older kids within this age group are probably roughly the same as they already spend, although now diversified and balanced among Growth Opportunities. The figures below suggest how the total time the child spends with interactive media increases with maturation and development. The proportion of time shifts gradually from Family Relationships to the other Growth Opportunities. The maximum online time increases as the child develops.

For further details and instructions, please refer to the recommendations and table and charts given in Chapter 3 on pages 36 – 37 and to www.mydigitalfamily.org.

EXAMPLES

Mom, Jon, and Cheryl have been playing a popular fantasy game for years. It started out much as fantasy about daily life when Cheryl was five and Jon was seven, but this game became a way Mom could spend enjoyable quality time with her children over ensuing years. They looked forward to sitting at the computer together and often talked about their gaming activities at other times.

Dad, who owns a successful business, Liz, seven, and Nathan, eight, play an entrepreneurial simulation game. The three arrange themselves around the computer with Dad in the middle. Each child has a special role and keeps a record on a pad of paper; for example, costs, profits, names, and locations. They have been playing for many months, and the game evolves through decisions they make together. They choose individual roles and profiles, but generally cooperate in running the business they built from scratch. The game offers good action, as they protect their transports from a variety of threats. Besides enjoying quality time together, Dad passes on his way of thinking in his work and teaches the children how to behave ethically.

Comment: In both games, families enjoy being together and also benefit from Socialization, Values Education, and Education Enrichment. See Chapter 14 for The Good Life, a game without violence that provides a balance of Growth Opportunities.

Who cares? Andrea is eight and has no particular interest in the Internet at home, except as needed for school work and to exchange e-mails. She is an exceptionally gifted musician and has studied the piano since age five. She spends at least four hours a day at the piano, but otherwise has a conventional school, social, and family life. Although Andrea's time is precious, her mother insisted that she spend at least one hour a week together with her online to see a variety of sites, including news, community, and her church group.

Eli is not interested. Eli is seven and already an exceptional athlete who plays sports every season. He spends no time online at home, except for doing schoolwork. Eli's dad tries to spend about two hours a week with Eli online.

Comment: Some kids want to devote any spare time and energy to develop a special gift or pursue a special interest. To them, the online home computer or other interactive media can feel irrelevant.

Weather: Is Dad OK? Seven-year-old Henry's father was stuck in Alabama on a business trip while tornados were passing through the area. Henry was frightened, so his older brother logged on to the National Weather Services' real time weather radar site. The brothers were relieved that their dad was not in the zone of the most severe storms.

Comment: Learning about weather, tracking thunderstorms in real time, understanding maps, appreciating distances and geography, gaining familiarity with concepts such as time and velocity, are all Education Enrichment learning opportunities possible from visiting such sites. Similarly, kids can enrich their current events and geography knowledge using online resources.

Guiding your kids. Seven-year-old Mary's mother logs on to Mary's portal whenever Mary wants to go online but lets Mary navigate within the site. Zach's mom lets him click on his personalized icon to access the site she selected for him. Both moms have told the children that they may not go to any other site.

Comment: Establishing rules about accessing online content cannot start too early. Give the child as many choices as possible for independent exploration within the rules you set and show him the possibilities a site can offer.

Shawna catches up. Six-year-old Shawna was having social difficulties at school. She was not interested in the same activities as many of her classmates. She would sit on the side of the playground and read. At home, she also enjoyed solitary reading and rarely went outside to play.

Shawna's reclusiveness caused her to miss important opportunities for companionship and social development. Her schoolmates interpreted her shyness as unfriendliness and avoided her, thus strengthening Shawna's avoidance and shy behavior.

Shawna was becoming more and more demoralized. The more she missed, the less she had in common with others, thus the more isolated she became. Her parents worried that she was caught in a vicious cycle and could be heading into serious problems. She told me in doll play that she felt unliked and was afraid of other children.

As part of her Media Plan, her parents encouraged Shawna to invite classmates over and utilize the online family computer as the main tool of interaction. Mom bought the latest software and made a special effort to find new and exciting sites that would promote interaction. As Mom helped and they explored, Shawna came out of her shell and made a new friend. The ice was now broken and Shawna began to feel more confident. Her natural charm emerged and she developed a good social life.

Comment: This is an instance of how interactive media play can provide Growth Opportunities for Socialization and Entertainment. Supervised therapeutic play unblocked a child's developmental trajectory. Some children develop more slowly than others, only to catch up later. This example illustrates how a mother used interactive media to remediate her daughter's relative developmental lag.

Obscenities. Eight-year-old Jill used swear words in instant messages she exchanged with Raquel, her sometimes best friend. Raquel had initiated a fight by insulting her, but Jill escalated by resorting to swears. Mother grounded Jill from going online. Mom tried to teach her about conflict resolution and the most effective and least destructive responses she might make. They discussed the use of expletives, alternate behaviors, and the impact words have on others.

Comment: The great variety of media interactions allow for many opportunities to observe your child. A mother sees her daughter's behavior, and takes the opportunity to apply appropriate standards to advance her development.

Molly needs room to grieve. Following her dad's death from cancer, eight-year-old Molly has been spending more and more solitary time in her room mostly roaming online, playing games, and, rarely, instant messaging her friends. Initially, her mom thought it was a reasonable way to stay connected to friends and to get their support during a difficult time, although she had the feeling that face-to-face support would be better.

Molly kept this behavior up for months despite her mother's gentle suggestions to give it up. It became a way of life for Molly, spending up to three hours a night instant messaging online, while claiming that she was collaborating on homework.

Mother guessed that Molly was withdrawing, but Molly denied anything was wrong. When her mother moved the computer to the family room on my advice, Molly continued to spend up to two hours online with her back turned to her family.

However, family members gently joined her for short times at the computer, and she gradually came out of her shell and became her old self. She responded well to sensitive efforts to reconnect with her family.

Comment: This is an instance of how a child initially used the online computer as a necessary safe haven during acute grief, but moved into using it as a hiding place from the challenges of relationships. Parents can often be helpful if they try to understand and intervene in their children's media life.

Where did I come from? Seven-year-old Peter spends online time with his grandfather learning about Grandpa's youth in the old country. They start with Google Earth and travel to other sites as Grandpa talks and teaches and Peter asks questions. They even find folk music they both enjoy and download.

Comment: Children can be curious about their connection with past generations and cultural heritage. Peter can allocate this activity to Growth Opportunities in family relations, Education Enrichment, and Entertainment.

Part 2: USER'S GUIDE

Congratulations! You can now make more of a positive difference to your baby's and young child's interactive media experience. Be sure to utilize safe online practices and filtering software. Even more importantly, make sure you approach your child with understanding, respect, and patience, and be as involved as possible.

In this section, we continue where the *Setup and Quick Start Guide* left off with a more thorough discussion of Values Education. Leading your family requires a fair amount of thought and planning, especially if you also have older children. You will think about safety from your family's perspective. You will see more thorough approaches to managing interactive media, even making meticulous gourmet menus for your child's daily experience.

- **Chapter 7:** More on Values Education
- **Chapter 8:** Leading Your Family
- **Chapter 9:** Safety and Other Challenges
- **Chapter 10:** Fancy Menus and Special Plans

Chapter 7: More on Values Education

I believe that teaching children values and ethical behavior is a *bona fide* and essential reason for bringing interactive media into their live. I hope to provide parents guidance that will stay relevant, even as technology brings our children gifts we cannot yet imagine.

Values guide our behavior towards others. The Golden Rule is the basis for any value system: We try to be good to others and avoid doing something hurtful. Ultimately, this function of our minds is rooted in our brains. Over the past decade or so, our understanding of how the brain works has began to accelerate.

The Brain, Mind, and Growth Opportunities. In a mature adult, almost seamless, complex calculations, often lasting mere milliseconds, and often operating outside our awareness, take place in and among brain centers to enable us to make the distinctions between right and wrong and to predict the effects of our behaviors on others. More complex processes take longer and operate within our awareness. We call the result of these brain activities 'judgment' and the rules we use 'values'.

Baby comes pre-wired with the potential to grow the hardware for doing these calculations. Starting in infancy, information processing in these neural circuits evolves into predictable patterns, as genetic blueprints interact with experience. The Golden Rule is the main lesson the child learns with increasing complexity as his brain matures.

We need to achieve internal balance among our wanting, needing, and doing in order to use the Golden Rule. These elements interrelate, evolving dramatically as a child matures. As a matter of fact, scientists are discovering that we very well might have a genetic tendency to learn such social rules and values, to do the right thing, or even to believe in a supreme being. At best, however, even the most mature adult struggles to find a reasonable balance, sometimes using quite a bit of brain power and emotional development—maturation of the right circuits—to develop such judgment. And no one ever does so completely.

To want and to know what you need is an essential part of being healthy, as long as it is balanced and in the right context. Without desire, we would not be motivated to do very much. Sometimes, the ability to want is sadly enfeebled. For example, desire lessens the more depressed a person gets. (I still remember a very depressed nine-year-old who had lost all desire and was not able to hope for any presents for his birthday.)

We call what the brain does collectively the "mind," just as we call the functions of muscles, joints, and bones "movement." The brain matures and grows, and the mind develops. Emotional and intellectual learning bring out of your baby's brain the biological potential your baby inherited from you. Development is necessary since a baby starts practically from scratch, mostly with a blueprint to develop her brain into an organ capable of providing a functioning mind.

Living in a reasonable family provides the normal baby with realistic opportunities to unfold and build all her organs from her genetic blueprints, including her brain. Hopefully, the baby eventually achieves whatever degree of balance, happiness, health, and a successful life with fellow human beings that these blueprints predict—or may even go beyond it.

But nature did not necessarily plan for, nor prepare a baby for many aspects of modern life. Humankind has created an enormously complex world for itself too quickly for biological adaptation. Successful interactions with technologies are part of our challenge now. Technology intersects with human development in children who interact with media.

So, when it comes to media, parents must select and feed young children a proper plan of Growth Opportunities, in this case, Values Education. A baby's brain, while resilient, is an extraordinary environment that requires incalculably the most sensitivity and care. From infancy, you are forming the core of your child's potential capacities, and as his brain matures, you will continue to enable him to build more sophisticated abilities on foundations he formed earlier. If you keep these facts in mind, you will have a good basis for managing your child's development.

This is how it starts in infancy: There is no distinction among self or other, nor awareness of differences among wanting, needing, or doing. Actually, there is as yet neither an "I" nor a "you." Using awkward language (and probably better: art, music, or empathy) we can only try to guess the infant's preverbal states: "Need now, want now, crave now, must have now, take now until content, maybe a feeding, maybe getting picked up, must have relief now. Nothing else will do or possibility even exists. Nothing else matters. Absolutely must get what want/need. If not, stay terrified that a major catastrophe is actually happening forever. All or nothing, no in-betweens."

Hardwired to stay alert to the baby's alarming cry, mothers automatically awaken to respond, even from a deep sleep. Eventually, baby starts making slightly different cries. Mother is hardwired to distinguish among them, responding differentially to less urgent pleas. Both baby and mother are now beginning to learn the difference in urgency between emergency life-sustaining need and less urgent want. Mother learns instinctively how long to wait to optimally frustrate or gratify her baby. All the while, the baby is evolving and establishing new brain structures to regulate his craving states which he will continue to refine over his lifetime.

One of the aims of Values Education is to teach the basic principle of human relationships, the Golden Rule. It implies the capacity to respect others—their wants, needs, and sensitivities—and to value the dignity of human life. It requires at least to distinguish between feeling and doing as well as between the self and the other's self. Because subtle variations like "Feel about others as you would have them feel about you" or "Treat others as they treat you" are not exactly the same.

EXERCISING THE GOLDEN RULE WITH A FRIEND

Actually, implementing the seemingly simple Golden Rule is a complex and challenging process, requiring enormous brain power, maturation, and the smooth coordination of many brain centers. A sophisticated brain is needed to implement this injunction.

Only humans have this enormously complex ability. These are parts of the process: Perceive a need; experience a desire; assessing your friend's condition, strengths and weaknesses; assess her similarities to you and differences from you; and predict what she might want; rehearse in your mind and re-rehearse your possible actions and their consequences, each time predicting how she might respond to each act and making small adjustments until you get a plan closest to right; assess likelihoods and suppress impulses that might injure her; maximize the chances that you would please her, drawing on memories and generalizations about specific experiences with her and generally with other people; draw on other stored knowledge of the world as it applies to the current circumstances; factor in appropriateness of time, place, and person; and assess her readiness. Then you repeat the above process and refine it until it all works together the best and feels right. Only then do you decide on what you would do. You assure yourself that you can trust yourself to act competently and that she will respond positively. Then you to pay attention to her reaction, and only then carry out your plan.

Your friend reacts, and in turn you react to her reaction, hopefully accurately interpreting her words, facial expressions, tone of voice, and other non-verbal behaviors—again drawing on enormously complex, highly developed discreet brain capacities of processing and evaluating sensory inputs. You experience feelings, perhaps of being more connected to her, or of increased self-esteem for doing a good thing, and then you might provide her with positive non-verbal and verbal feedback. You might also learn something new about her or yourself. Or maybe she disappoints or even hurts you when her reaction surprises you or is too different from your prediction. You then evaluate your inner reaction to her reaction and test it for appropriateness and accuracy before actually acting on it.

Simultaneously, you observe and add this experience in whole or part to your accumulating stores of memories. Modifying other brain centers that will include your images of this friend and yourself may lead you to generate rules for improved judgment in future interactions with this friend and others. All of this is choreographed, occurring spontaneously and quickly, mostly outside your attention to it. If you have capacity for insight, you can choose to reconstruct what just happened and learn even more from your experience.

If you were hurt, you might suppress the whole thing and distract yourself as a way to cope. Or you might replay the scenario over and over in your mind, trying to alter one step or another so that you could understand what happened and learn to achieve better future outcomes. You might also draw on abilities to comfort yourself or seek comfort from others, or in other ways, even through prayer. Or you might forgive this friend. Or, if you feel safe enough and have the skill and courage, you might speak directly to her about what just happened to repair the relationship. Or you might have some kind of tantrum and act in destructive ways against yourself or her or others. Or you might also work to manage your angers and fears without acting destructively.

People differ in how well developed they are in the various aspects of this process and how they react. Some tendencies are inherited; others learned. Many retain some infantile characteristics. Still others are more highly developed. Fortunately, most people are developed well enough that they do not act destructively and get a fair amount of guidance from their moral compasses.

In your baby's early life, how you sense and respond to her needs is what really matters. Basic trust that others are good and that you are effective in getting them to respond to what you have to give form an important core for any moral life. When a toddler begins to experience himself as "I," and others as "you" (which is initially a rudimentary and dim clone of "I"), he also develops an insistent and stubborn "I want/need now" in the entitled and urgent manner of the infant. His mobility and potential for mischief can make him quite a handful.

With the development of language, especially by age three, parents can begin to make distinctions for the child between wanting and needing because he is ready to learn and eager to please. For example, Jenny's mother makes this distinction as Jenny begins to tantrum in the supermarket: "Jenny wants this box of cereal. She thinks it would be nice and make her happy. But Mommy will take care of Jenny, so that, even if she doesn't get it now, she will feel better and nothing bad will happen to her. Jenny does not need this box now. She needs hugs more," or words to that effect. Eventually, Jenny will learn to think this way for herself. Learning the distinction between needing and wanting, to wait, and to be patient gives the child a wide range of tools to live a better life. For the older child, discussing daily the moral aspects of what happened that day in school or on the playground is a wonderful practice.

Because these processes are so intricate and depend on so many factors, there can be problems with the balance in the mix of wanting, needing, doing, and thinking of others, including impaired basic brain functions or relationships that interfere with early development. Poorly regulated individuals can be destructive when succumbing to extremes of passion, rage, impulsivity, and/or undervaluing of others and of human life.

An urgent desire or need can often find a substitute in something else more accessible. A desire can be so strong that it feels as urgent and as powerful as a need. Blurring this distinction is a main technique of merchandisers, so that buyers come to feel and think of wanting is as compelling as needing.

Bottom line: Values, translated into a sense of what is right and wrong, good and bad, are essential for achieving balance in the mix of wanting, needing, and doing when we interact with others.

PARENTING BASICS

MAKE A SAFER WORLD: RAISE DECENT AND CIVILIZED CHILDREN

Online safety is closely tied to prevailing societal and cultural attitudes about decency in human relationships. These include general issues of social responsibility, exploitation of children, violence against children, violence in the media, and violence in general. Although dishonesty, greed, and violence have always been part of human life, parents can do a great deal to assure a culture of peace, decency, and tolerance by how they raise children. So we start at home.

Discuss with your partner, and eventually children, ways to keep moral behavior in the forefront of your family life. Decide together with your partner on the values that matter to you and what you want to convey to your child. Your child will look to you for clarity, so know in your own mind where you stand on violence and how you employ your beliefs in daily life.

Probably rooted in the biology of the brain, moral development evolves cumulatively like other aspects of the mind. Parents who take seriously their role must regularly review their own routine practices at home in conveying correct values.

Your baby is already learning essential core attitudes of trust and security directly from how well she has been treated from early infancy onward. Make sure you understand thoroughly your baby's unique developing physical and emotional needs, meeting them as fully as you can. Never direct violence—verbal or physical—at your child and always shield the child from witnessing violence. If you show your child that the world is an unpredictable or dangerous place and that violence is acceptable, she will have a hard time unlearning these lessons later on and may never succeed at that. Very young children unaware, adjust their own behavior by imitating parents and defining what is right and wrong strictly by what it takes to please their parents and avoid punishment.

Early in the preschool years, you can begin teaching your child kindness, personal responsibility, compassion, positive problem solving, respect for herself and others, the Golden Rule, the value of individual life and community, as well as the balance between self assertion and social responsibility. It is better for parents to be proactive than reactive. Teach and model such values throughout your child's life. Avoid exposure to media violence, including cartoons and news.

Early school-age kids are more aware of right and wrong as they become adherents to a rigid system of consistent rules and punishments. Later, school-age children begin to form internal moral compasses and to experience guilt. In later stages, moral ideas and behaviors become more and more refined as kids are introduced to abstract thinking and social sensitivities, enabling finer distinctions within moral grey areas.

Children welcome help as they struggle to figure out what is right, what is wrong, and the gray areas in between. Always ask the child for his opinions as a starting point. Keep the conversation short and simple with younger children and use examples from fairy tales and stories and play with dolls and pets. As he gets older and begins school, engage your son or daughter in active conversations about moral and ethical matters whenever opportunities present themselves. Review moral matters with the child daily.

Values that are not practiced are meaningless. Encourage behaviors (in virtual as well as everyday reality) which clearly manifest taking personal responsibility for mistakes, kindness, fairness, honesty, respect, compassion, and the just treatment of others; or advocacy and actions for issues currently discussed at school, in church, or that address current events in the community and beyond. Of course, your own leading by example must underlie any approach.

Children benefit from learning that anger is a common and normal human feeling. But they must learn the difference between feeling angry and acting aggressively, a lesson that should be repeated often. Learning to label the feeling and to resist the impulse to act on it takes much practice. It is also important to teach children to expect that conflicts and differences will normally occur between well-meaning people but must be resolved without violence.

Competitiveness and assertiveness based on respect for self and others are healthy, while bullying or being bullied are not. Teach nonviolent ways of expressing and channeling anger and frustration. Children must learn that violence has serious consequences for both victim and perpetrator.

Teach older children and preteens that impaired judgment caused by alcohol and other drugs often leads to violence.

Teach that mental illness is not a cause of violence, and violent people are not necessarily "crazy" or mentally ill. For example Sadaam Hussein, Josef Stalin, or Hitler were perfectly sane men who were highly successful in what they tried to accomplish, but they were evil. Also, teach older children and teens that there may be times when violence can be an acceptable solution (e.g. war, self-defense, preventing terrorism), but always with restraint, only as a last resort, and never when driven only by impulse or personal anger. Help the teen channel aggression and energy away from violence to high risk behaviors and into positive and constructive activities instead.

When a child is aggressive or violent, try to understand the cause and give guidance and appropriate discipline or punishment. Ask the child to think about what problem his violence was supposed to address, and help him think through alternative solutions. If he does not respond to your efforts, obtain professional help rather than getting into an endless spiral of negative interactions. Some children, like those with Attention Deficit Disorder, find it hard to learn from experience.

Raise moral or ethical questions with your kids based on what you see in the community or elsewhere. Discuss films and books in terms of your values. Do your best—there seems to be insufficient glamour in moral or ethical behavior to most kids, so finding opportunities that compete with the razzle-dazzle of popular culture is always a challenge.

Interactive media provide opportunities for values development that includes learning and practicing considerate behaviors toward others. Searching sites showing human tragedy like a natural disaster, suffering from war, or genocide like the Holocaust and Darfur, finding your community's volunteer and social agencies sites, and linking to your own or another religion's site to see how each pursues moral and ethical causes can provide these opportunities. *The Good Life,* in Part 3, is an example of a wished-for future action fantasy video game that promotes Values Education without violence.

Two ways to inculcate values and enrich and strengthen your personal, family, and community life are to involve your family in community (including virtual) or community service projects, and encouraging each family member to perform at least one good deed daily. Review these together as a family at dinner.

Most importantly, your own actions really matter. Try to put into practice nonviolent values in your own life (with help from your religion, if that is part of your life). Refrain from directing physical or verbal violence at others. Your own behavior—personal responsibility, compassion, positive problem solving, respect for yourself and others, practice of the Golden Rule, respect for the value of individual life, and a striving for balance between self-assertion and social compliance—especially as practiced toward all in the household (including the child, siblings, grandparents, and even pets) is your main means of teaching such attitudes.

WHERE HAVE ALL THE HEROES GONE?

Beginning in preschool, children are eager for heroes. They have always looked for inspiring heroes whose prowess in action, deed, or word they can emulate. They want to learn principles of correct behavior and can be introduced to ideals of courage, faith, hard work, perseverance, endurance, selflessness, idealism, love, community, and the pursuit of knowledge, beauty, and freedom that go beyond the daily lives of their family. Families used to teach the importance of heroes, often even displaying their photos on the wall (e.g. FDR, Einstein, Jackie Robinson, Martin Luther King, etc.), and kids welcomed such powerful influences. But today's kids are mostly left to adopt, at best, shallow celebrities, greedy entrepreneurs, or antisocial animated action figures.

I believe that bringing in and keeping heroes alive in your home is a far better way to turn out a future Einstein, Mozart, Salk, King, or whoever else best exemplifies your ideals. Deciding who your own heroes are and introduce them into your home by regularly admiring and celebrating their achievements and keeping a conversation going about what it takes to be such a hero. Highlight the idea that real people are authentic heroes and discuss how. Interactive media, including videos on (and see Chapter 14 for a video game), and a wealth of text and images online offer opportunities to bring the images and, words of admired people, who can inspire children toward the best values and the best of what it means to be human.

PARENTING BASICS
TEACHING VALUES: COURAGE

Courage is commonly recognized by nations as selfless sacrifice in battle. In some cultures competitive, courageous or violent acts mark passage into adulthood. And there is also the lone hero kind of courage. Courage is a quality individuals are not supposed to claim for themselves. We are supposed to wait to be recognized and admired by others for our courage. "I am brave" or "I have courage" sounds immodest and too much like bragging. Who ever heard of getting a medal from yourself for courage?

But there are different kinds of courage. In all, courage enables a deliberate and persistent striving toward a goal (generally regarded as good) that calls for knowingly facing difficult challenges (pain, fear, danger, or other difficulty). Courage is not reserved for military heroes. It can be highly personal and can often go unrecognized by others.

Countless people live lives of quiet personal courage as they face or recover from adversity, disability, injury, loss, illness, or violence. For people suffering from unseen tortures like post-traumatic stress disorder or depression, or recovery from addictions, just getting through the day can take enormous courage. A life shattered by violence takes considerable courage to repair and rebuild. As a nation, we are living through difficult times that call for courage from so many of us, including our leaders.

Ordinary people show courage when they push themselves and persistently strive (despite shame, self-doubt, pain, fear, or temptations to take the easy way) to do something good and helpful. Teach your child that we can all be courageous and that we can all support the courage of others. Spirituality, community, and faith help us find and hold on to courage.

Challenge other common beliefs and practices: Teach your sons or daughters that courage is special but not rare and that he or she can and should strive for it. Teach your child that to be courageous is not to be without fear—it takes no courage to do something when you do not fear it. It is not courageous to take foolish risks, react thoughtlessly or impulsively, or act violently.

But most of all, teach your child to claim personal courage for himself. Teach your child that courage is within reach and not reserved only for others. Look at your child's daily accomplishments and teach her to appreciate her own unique kind of courage. Look for nonviolent strivings toward a personal or social good that show persistence in spite of pain, self-doubt, fear, or shame.

Encourage your child—enable him to know his own courage. Courage is episodic—we can be courageous about some things and not others. Your child can be brave even when he cries from pain or hesitates from fear. He can be a hero just for trying as hard as he can. Say this to your child: I am proud of you. You are brave.

Challenge the conventional practice of waiting for others to recognize courage. Teach your child to say this: I claim courage for myself. I claim a medal for myself. I am brave. I am courageous. I am a hero.

And remind your child to be grateful for the blessing of courage and for others who enable it, because courage is often collaborative.

While interactive media can help promote a culture of decency and peace, such content is pretty much absent from the available activities that consume so much of our children's time and attention, especially as they grow older. You may find online content specially directed at ethical, moral, or spiritual guidance, for example church-related sites or charities like the Red Cross, to advance moral development and social responsibility. Civic involvement and political awareness are values that you can promote by introducing appropriate sites to young children. Spiritual and religious, social activism; local, domestic, world awareness, or social issues sites can promote awareness needed by adolescents for healthy development. The school-age child can be introduced to these resources in your presence.

Chapter 8: Leading Your Family

I am devoting a chapter to this topic because leading your family well will only enhance your effectiveness in implementing the Media Plan as well as provide more general benefits.

How you introduce new ideas and routines into your home is always a challenge, especially when children are older. The first and most obvious step is to determine if the effort is worth it. When partners work together, they model collaboration and teamwork to their kids. When they take the lead, thoughtfully planning and making changes together, they model problem solving and leadership. To help parents find constructive ways to integrate the Media Plans for all the children, I provide a peek into my recommendations for older kids.

Craft your role. Think of yourself as a leader who is motivating and bringing others along for an important collaborative lesson. Your first resources are the people who live with you and your child(ren). Don't forget grandparents or other relatives in the household or nearby. Bring in as many family members as stake-holders and keep them involved, especially your spouse, partner, or other important adult in the home. You will be much more effective as a team.

Show other team members the menus you prepared and build consensus. Be patient. How you approach your children will determine your success and may even be an opportunity to generally strengthen your collaboration as a family going forward. I suggest you introduce every new step of the process as a joint project whenever possible. Continually reassess whether your efforts are getting the results you want.

Obviously, much of this depends on your child's response to this process, so pay attention. You are empowering others to think together, collaborate, cooperate, and achieve a family consensus. Your older children will show you around their media world. These can be set up as favorites in their Media Plans.

To lead effectively, you must know commit yourself to understand the technologies in your home. As when teaching kids to drive, parents become knowledgeable drivers themselves. There is no shortcut here: if needed, take an evening course in your local high school or junior college, or teach yourself from books or an online course. The more you all know about the technology of the computer, Internet and other media, the safer you will be.

PARENTING BASICS
APOLOGY AND FORGIVENESS

As hard as they try, most parents "lose it" occasionally. A parent who rarely mistreats a child can repair damage and promote development with a full, open, and dignified apology. Remember, the reality is that this is not a level playing field—you are the leader and the child is dependent on you.

Recognize and acknowledge your child's injury to him ("I know I hurt you. It wasn't right. You don't deserve to be hurt this way," not "But you hurt me too.")

Take genuine and direct responsibility ("I was definitely wrong!" not "I didn't know what I was doing," or "You made me do it.")

Validate the child's legitimate right to a reaction ("You are right to blame me and be angry with me," not "Please don't be angry with me.")

Apologize sincerely and mean it ("I am really sorry!" not "You should be sorry too because it was partially your fault.")

Ask for forgiveness ("Please forgive me," not "Let's forgive each other.")

Resolve not to repeat this mistake and make a promise you can keep ("I will really try not to do this again," not "I'll never do it again.")

Respect the child's need to think and decide ("Take your time to think about whether you can decide to forgive me. It is important to me that you really mean it if you do.") Seek positive cooperative and collaborative opportunities with the child.

Empower the child with a real choice. Do not pressure or bribe or expect automatic forgiveness ("If you don't want to forgive me, I will understand.")

Accept forgiveness with gratitude ("Thank you. It really means a lot to me that you forgave me.")

Graciously accept the child's hesitation or refusal to forgive you without threats of retribution ("I will ask you again later because your forgiveness is so important to me.")

Refrain from your damaging behavior in the future. The child can then feel respected, learn that injury can be repaired and that her own forgiveness can be powerful and good; that occasional anger can be part of good relationships; that people can be trusted, even when not perfect; and that it is good to take responsibility. These are valuable lessons about how love works in human relationships.

PARENTING BASICS
IMPLEMENTING A NEW IDEA IS FAMILY BUSINESS

Use every opportunity to demonstrate and teach respect, kindness, and patience to your children.

Line up your ducks.
- Decide on your own role.
- Become interactive-media savvy.
- Make sure your co-parent is on the same page.
- Support each other and clarify your roles.
- Decide on how to proceed and how much flexibility you will show.
- Include children's preferences in the plans.
- Design your proposed preliminary plan and daily menus

Be a negotiator and problem-solver.
- Announce your general intent.
- Explain, negotiate, and set the rules of the coming debate.
- Respect differences—agree to disagree.
- Maintain order firmly but gently.
- Listen and ask open-ended questions.
- Help your child explain clearly his viewpoint and goals for himself.
- Sit next to your child while he is online, take the role of a respectful pupil, and ask him to show and tell, take you around, and teach you.

Build your team and lead by consensus first.
- Help your child make sense of your goals and plans in her own way and at his own pace.
- Keep the conversation going over a period of weeks. The more you include your child in the legwork, planning, and implementation, the more he will cooperate and learn to collaborate.
- Ask the child to show you the media activities he has already discovered and incorporate these as much as possible into the new menu.
- Include your child's suggestions and ideas in the project whenever possible.
- Enlist your older child to help with his younger siblings or other relatives. Be flexible and experiment together.
- Include the entire family and respect each member's contribution.
    - activities, rewards, and consequences.

Lead without consensus, if necessary.
- Announce that you prefer cooperation and enjoy relationships, but that you will move forward with or without cooperation.
- Be increasingly firm.
- Make announcements about consequences for noncompliance.
- Provide a timetable and clear limits and conditions.
- Make sure that the child understands her opportunity to influence the plan will not last indefinitely. If she blows it, she will live with the consequences, and you will not take the blame for her bad choices.

Final steps.
- Decide with your co-parent on the final plan and timetable.
- Gradually shape the time and content of your child's Internet use and other forms of interactive media.
- If the child had been spending too much time with interactive media, reduce this time gradually over a period of weeks.
- Expect that your child might test limits and be prepared to follow through on consequences.
- Review, adjust, and renegotiate regularly.

Tact and team building. The facts of good nutrition are not debatable, but how, when, and where meals are prepared can be. Keep asserting that nutritional goals are basically not negotiable, the family can be creative in how they are accomplished.

Remember, no matter how you decide to structure your children's interactive media time, you are the leader of this project. Set up special family meetings to discuss your project. Describe what you've done, what you still need to do, whom you have talked to, and the reasons you are doing all this. Especially when installing new software or hardware, announce that the new plan also increases the power of the family computer to benefit the family and children.

MEDIA	**FOOD**
No parent involvement Mostly games, unproductive Missed opportunities Safety Risks Other Risks	*No parent involvement* Mostly junk foods Poor nutrition Health risks and obesity
Minimal parent involvement Minimal parent involvement Monitor child's online practices and encourage a balance of Growth Opportunities	*Minimal parent involvement* Minimal parent involvement Monitor child's eating practices and encourage a balanced diet of food groups
Parent(s) more Involved Limit media time to activities offering Growth Opportunities **Have family media session daily** **Balanced Media Plan** Better Family Relationships Healthier Socialization Sounder Values Education More Education Enrichment Less Entertainment *Better/safe nourishment* Better family life Better use of time Emotional and intellectual development safe from external influences Healthy media habits and discipline	***Parent(s) more involved*** Limit intake and provide only healthy choices **Have at least one family meal daily** **Healthy Diet** High fiber and pasta products Fish, eggs Fruits and vegetables Dairy Less Fat *Better/safe nourishment* Better family Life Physical growth and health Brain growth Healthy nutrition and fitness

Do not hurry through this phase. Be flexible—old habits die hard, even for kids. But gradually tighten up your expectations, so that by the second or third week, your plan is fully in effect as the sole interactive media exposure your child will be allowed. Agree to a transition period from old habits to allow older children and other adults to get used to the new arrangement as well as to see what works and what does not. You may want to keep a log and have your child keep a log of interactive media time and uses, using this baseline as your starting point. You may need to do this over several weeks and modify your goals as you go along.

A good leader builds consensus and motivates. So be sure to listen to your children and empower them. Make the planning and shopping a family project that includes everyone. Making your children stakeholders in the project will increase their motivation and success. Be creative and devise new ways to help the family have fun and learn together. Don't be afraid to explore and change things as you go along. Most importantly, rely on your own common sense and flexibility to get the job done. However, if you do not get very far, you might have to eventually call on your authority as a parent to carry out your duties and responsibilities to look after your family's health and well-being.

PARENTING BASICS
IS YOUR FAMILY SAFE FOR KIDS?
GUIDING, DISCIPLINING, ABUSING, AND APOLOGIZING

Good parents are not born: they are self-made and are always working hard to remake themselves. Since I am providing guidelines that may put you into difficult situations, I want to remind you how to react in the most effective and constructive manner. We can all always use a reminder of what is best at such times and relearn what it means to apologize.

WALKING THE LINE: GUIDANCE AND DISCIPLINE

It is always better to reward good behavior. Be alert to recognizing such behavior—or the absence of bad behavior—and reward generously. Parents can plan together with the child a hierarchy of rewards and post these on the refrigerator. Always remind the child, and yourself, that mistakes happen and we all must learn from them.

A child's misbehaviors or defiance, your responses, and the transactions and mutual learning that follow, are all normal parts of growing up, starting with the "terrible twos" when the child begins to actively explore his world. Although adolescents can make this task difficult because they are also trying to experiment with independence, these principles remain true. The more time and effort you invest building, thinking, planning, and maintaining, the better.

Teach your child that he makes a choice about every action, that there are consequences to every behavior, and that he has an ability to influence the outcome through the decision he makes. Teach your child that not thinking is no excuse when it comes to responsibility for misdeeds. Teach your child that his behavior can narrow the options you have in responding. Teach him that his decisions and misbehaviors can be so powerful that they leave you no option other than punishing him.

This learning best occurs within a preexisting, positive relationship with a respected and loved parent. In the long run, a child has more to lose when facing disapproval by someone he loves and respects, than from fear.

It is best to avoid either extreme—ignoring misbehavior or being overly punitive. As the child develops, he hopefully will learn from prior disciplinary episodes that while in the short run, he may experience fear, in the long run, he will recover, and in no way will he face abandonment or rejection. In this manner, the disciplinary action is like an inoculation: although it causes some mild pain, it is actually building health in the long run.

Before disciplining your child, review your own role as a leader. How did you teach the child to make the decision to behave in a certain way? Have you followed through consistently? Are you modeling proper behavior for the child? Are your expectations realistic? It is best to supervise a child rather than leave him in situations where he may predictably face a peril, or misbehave because of poor judgment or ignorance on your part. Avoid punishing a child because you have placed him in a difficult situation.

Next, before discipline, there is guidance. Try to regard the child's misbehavior as a mistake in judgment due to ignorance. If you are too upset to see it this way, back away and return when you can. Under your leadership, children learn best in helpful and supportive collaborations. They need repetition. Try to see misbehavior initially positively—as misdirected efforts at learning—and react by teaching rather than disciplining. Ask him what his goals were and teach him other ways to achieve the same goals. Make apologizing part of the penance.

Being firm and keeping discipline within a loving relationship is difficult. Most parents dislike disciplining because it may cause their child pain or fear. They realize that the child gives them his most precious gift—love and trust—and don't treat that lightly. However, parents must realize that striving toward self-control and proper behavior is a lifelong process, and no single episode of punishment will accomplish this goal.

Even when guidance fails, the purpose of discipline and punishment remains to educate and promote development. Even when a behavior is so egregious that its recurrence must be prevented at all cost, a principal purpose of discipline is to teach self-control and correct behavior. All children need discipline to learn how to accept responsibility for their actions. Additionally, discipline is necessary to assure a child that he is safe from danger in a confusing and scary world because someone loving and responsible who knows more is in charge.

Mete out discipline in a manner that would optimize learning proper behavior. Match the punishment to the behavior fairly, telling the child how you arrived at your decision. Striving to be respectful and fair, make sure your response matches the child's ability to comprehend and learn from it. Add that while discipline is a necessary consequence for a behavior, the child himself is still a good person who will learn to be even better. When a child fesses up readily, show your appreciation for this integrity and soften—but do not eliminate—the consequence. On the other hand, when he covers up, make the consequence harsher or add another.

Discipline works best when it is appropriate to a child's understanding, rationed in a manner that does not permanently injure physically or emotionally, and is not beyond his ability to use it for learning. In the same vein, it works well when it is consistent and measured in intensity to the offense and the child's development. A system of escalating hierarchies of consequences, starting with a warning, followed by milder and then more severe punishment is best. Start teaching children the hierarchy of misbehaviors and consequences so that they know what to expect. Post the hierarchy on the refrigerator along with rewards for good behaviors. Follow through consistently.

It is best when discipline is shared by both parents. Work together and avoid letting the child drive a wedge between you. Agree on and anticipate the transgressions that deserve discipline, and organize these in order of seriousness. Focus on one or two at a time. List all possible punishments available to you in order of severity. Agree that when misbehavior occurs, you will each start with a selected punishment and increase the level of severity if the child does not comply. Back each other up. Be progressively stricter with second or third offenses. Be consistent and accept no dishonest reasons or excuses.

It can be a challenge to figure out appropriate punishment. Inventing a system of discipline during the heat of the moment is impossible. When you are angry, you can forget that you are a big and powerful adult while the child, no matter how difficult or frustrating, is small, not as capable, and dependent on you. Your child is fundamentally weaker than you, needs you, and does not need to be humiliated by your excessive show of power. Your child is a person. To injure, to demean, or to humiliate is to abuse. These are forms of verbal violence which is just as scarring as physical abuse. These have no place in childrearing and can distort development of the child's brain.

Knowing how the child's school and other caretakers apply discipline, adjust all methods so they are consistent . It is best for the child that two parents work together in agreeing ahead of time about thresholds and hierarchies. Revising the hierarchy together and reviewing it with the child is best. In some instances, the child can take part in negotiating limits and punishments.

THE CHALLENGE

Some children continue to exhibit troublesome behaviors no matter what you do. Children who respond poorly to discipline—with excessive defiance or compliance, or inability to learn—would benefit from a professional evaluation before their behavior permanently damages relationships with their parents and begins to erode the child's self-image.

Typically, the child's behavior catches you off guard and floods you with feelings of impotence, disappointment, anger, or even rage. This moment is one of the most challenging of parenthood. It is normal to become impatient and frustrated with children at times. However, you must avoid by any means injuring a child. This is not the moment to react. Walk away, count to ten, say, "I am too angry right now. I will deal with you later." Take deep breaths, walk out of the room, wash your face with cold water, take a shower, vacuum or do the laundry, shop, pay bills, or do sit-ups, whatever you need to do to calm down and get it together. Feeling angry does not mean acting angry. That is an important distinction to teach and re-teach your child by example.

In addition to not punishing a child when you are angry or frustrated, also avoid doing so while impaired by fatigue, alcohol, or drugs. You could too easily lose control and direct naked aggression toward the child and abuse him. This goes for both verbal and physical aggression. Punishment is never an occasion to vent frustration. "I am only doing this for your own good ... One day you will thank me ... You will know I did the right thing when you have children yourself," are often thinly veiled excuses for your loss of control. Attacks against children are profound betrayals of trust and love that wound deeply, long after any physical injuries heal. Brothers and sisters witnessing such aggression can be similarly damaged.

The best way to solve a problem is with a clear mind as you strive to think clearly and plan ahead before punishing a child. Co-parents could plan together, back each other up, and remain consistent.

These are the questions I strongly suggest you ask and answer before disciplining or punishing:

What will I accomplish by lashing out?
Is this a good time to punish my child or am I too angry right now?
Do I really want to hurt my child emotionally or physically and damage our love for each other?
What can I do to cool off?
What is the best way to have the child learn what I want him to learn?

You could seek out another grown-up or call a hot line to discuss and unload. A bystander, co-parent, or another family member could offer such help, too.

All children have a right to safety. Bystanders witnessing abuse should intervene to protect the child and assist the parent in regaining control. If a parent repeatedly loses control or attacks a child physically or verbally, she should be offered professional assistance. In the meantime, the case should be reported to appropriate authorities, because the child has a right to protection.

SPANKING

Spanking has been a commonly practiced means of disciplining children in many cultures: "Spare the rod, spoil the child." However, countless children have been raised well without a single spanking. Moreover, I do not condone spanking because the line between spanking and abuse can be too thin for many parents.

Parents have the awesome responsibility of raising the next generation of civilized people. Child development experts know that children turn out best when encouraged with verbal rewards for desired behaviors and rewarded for the absence of undesirable behaviors.

Parents should strive to gradually shape children's behavior in a context of nurturing, love, and respect for the child. Children can also be invited to join in a discussion of what would be a fair consequence to empower them and teach them self-discipline and fairness.

BE ALERT

Abusive parents are often under great stress themselves and need support. They may have serious emotional problems. They may lack skills, material or emotional resources necessary for raising children and may need assistance from a family service agency or other family members.

Children who do not respond to reasonable rewards or punishments may have learning problems, attention deficit disorder, or other difficulties. Such children are more likely to be abused because they can lead even the most patient parents to feel frustrated, guilty, ineffective, and unrewarded. Parents or concerned family members may obtain support from a professional.

BEDROCK PRINCIPLE: THE FAMILY—FIRST, LAST, AND ALWAYS

We already know much about human development and relationships that can be applied to help our babies blossom into fully-evolved people. I urge that any future application of interactive family media include support of families and children.

We will need to be vigilant to assure that these principles are implemented, even as commercial, economic, market, and political forces shape the uses of technology. We must advocate for adherence to well-established principles already practiced by enlightened family support agencies nationwide. Our hopes for a free and humane world rest squarely on how we educate and raise our children.

IS YOUR FAMILY A SAFE PLACE?

My entire approach rests on building and affirming strong Family Relationships. The Media Plan can succeed only if your family is itself safe. The advice here is not intended as a quick fix for broken or impaired relationships or families, but only for reasonably healthy families and well functioning parent-child collaborations, and only with the intent of giving parents ideas of how to improve them. To use my suggestions, parents must be at their best when interacting with their children and keep their children safe. That means that the parent is in a good mood, in a teaching and learning mode, fully attentive, patient, available, and absolutely not under the influence of alcohol or any other substances that may impair emotional response or judgment.

But not everything will work for everyone or work in the same way for all families. The strengths and weaknesses of your family life and parent-child relationships are the most powerful predictors of how well you will succeed in making any new change. Habitually communicating together flexibly and effectively, agreeing to disagree, having clear role definitions and intergenerational boundaries, expressing mutual respect, and using already established working ways of problem solving together are the most important predictors of how well the child will cooperate with any efforts you make in any area, including media use.

Generally, this means that family members habitually collaborate in setting up and benefiting from common projects such as vacations, redecorating rooms, household chores, choosing Entertainment options, or using the family car. Family members are more helpful than disruptive and enjoy each other's company. Parents guide and set limits and children respond reasonably well. Family members appreciate their small community and pull their weight, more with cooperation than friction. When conflicts arise—and they always do—things get worked out reasonably well and without lasting resentment. It is best for all concerned that both parents are on the same page. More commonly, one parent takes the initiative. Sometimes the other shows little or no interest, or willingly defers to the spouse. Usually, making a sincere effort to acknowledge, accept, or patch honest differences and agreeing on roles can be sufficient. If this sounds like your family, you are likely to get more benefits than problems from the Media Plan.

Unfortunately, some families find only limited comfort together because feelings run deep and hard, or don't run at all. Sometimes, these families find a stable way to live together in such circumstances, knowing what to expect from one another. Sometimes family members form alliances or seek emotional distances from each other in order to feel safe and avoid conflict. Sometimes a co-parent may go along reluctantly, harboring resentments, finding fault, or generally obstructing the project. Such a development may signal that the project has become a lightning rod for serious underlying family problems.

The fit between any one parent and child is rarely ideal. In most cases, parents initially work hard to make the relationship work, and later the more mature child joins in. This effort is an important task in the psychological growth of both children and parents. However, in some sad situations, the differences in temperament are too great. I have seen intelligent, well-meaning parents who are mismatched with one or more of their kids. While infancy may seem to go well, friction and behavior problems begin to worsen as the child grows and develops her own personality.

For example, a restrained, well-organized, and ambitious mother discovers her five-year-old son to be learning disabled, inattentive, impulsive, and aggressive. The more she tries to change him, the worse their relationship becomes. Usually, the child cannot make the changes that would work, and often feels less loved than his "better" siblings, growing up with a sense of "badness." He has less and less to lose as mother-child love is buried under heaps of anger and distrust. Ideally, this parent could let go of her disappointment, and even shame, and accept the child for all the good in him.

However, it's sometimes impossible to accomplish this difficult "gut level" adjustment. It's also difficult for any parent to accept the reality that the child will not become her version of the best he can be. Parents need the gratification of feeling they are doing a good job; such a child cannot provide enough. She may be embarrassed about his behavior. Feeling like a failure, guilty, ashamed, and angry, she is in pain. Certainly, it helps for a co-parent to become the main parenting person, but this is not always possible. These are difficult situations that have no perfect solutions and may require outside help. Not infrequently, interactive media become lightning rods for the conflicts that arise from such incompatibilities.

These and other families may not be able to enjoy the benefits offered in this book because it is unlikely that the online and interactive media experience will be any more rewarding, less painful, or essentially different than their usual experiences. When there are ongoing or escalating problems around the family computer, it is usually because it has become another lightning rod for severe existing conflicts.

Sadly, for some families none of my guidelines will work. In such circumstances, parents may face several choices: Parents may accept conflict and tension and live with it; parents might have to settle for a lot less than they want, capitulate to the children and choose other battles; or parents may choose to eliminate all interactive media from the home without negotiation. The best outcome for many families is this: when conflict around the uses of interactive media causes so much tension, family members become motivated to obtain professional or pastoral help.

In new or young families, the online and interactive media experience may unmask problems that lie just under the surface and have not yet emerged until a challenge is placed on the relationship. People may show impatience, tendencies to dominate, excessive criticism or competitiveness, excessive sensitivity, unreasonable or inappropriate behavior, or raw conflict with one another. Children may be uncontrollable, hard to engage, or inconsolable. On the whole, the interactive media experience may seem to be more trouble than it's worth. Parents in such families would do well to find a way to address these telltale signs. A positive first step is to realize honestly that they face challenges in their family that require help.

Although sound principles of child development underlie the easy language in this book, it is not intended to provide of professional diagnosis or therapy for any specific situation a person or family may face. In fact, seeking such help from sources other than a professional who would conduct a thorough, personal, face-to-face consultation is generally unwise, especially when it comes to children and especially online, because the Internet is too full of bad advice, testimonials, and misinformation.

WHAT'S AHEAD AS MY FAMILY MATURES?

You may find this preview helpful as you continue to manage your Media Plan. I want to introduce you to a quick peek at what I would recommend for your child as he grows past the ages covered in this book, so that you can put the entire system in a larger context. Additionally, the trend with interactive media is to relentlessly move applications from older to younger children, so looking at these older kids will clue you in on what to expect. Moreover, you may find the specific guidelines to use with older kids and teens in your home.

The same general principles apply as for younger kids. However, please notice how total time and alone time increase, as do proportions of Growth Opportunities. Family Relationships remain a significant part. If you want to include all media, you can double and even triple the times given, but try to approximate the proportions.

Hours /Week (Digital)	Age 2	Age 5	Age 8	Age 12	Age 17
minimum Family Relationships	1:00	2:00	2:30	2:30	2:00
maximum Socialization			:30	1:00	2:00
minimum Values Education		:30	1:00	1:00	1:30
minimum Education Enrichment		30	:45	1:30	1:45
maximum Entertainment			:15	1:00	:45
total maximum	1:00	3:00	5:00	7:00	8:00
total minimum			1:00	2:00	2:00
maximum per session	:15	:45	1:00	1:30	1:30
minimum with parent	1:00	2:45	2:45	2:00	1:00
minimum parent nearby		:15	2:00	3:30	2:00
maximum independent			:15	1:30	5:00
+ Hours /Week Other Media	3:00	6:00	10:00	14:00	16:00

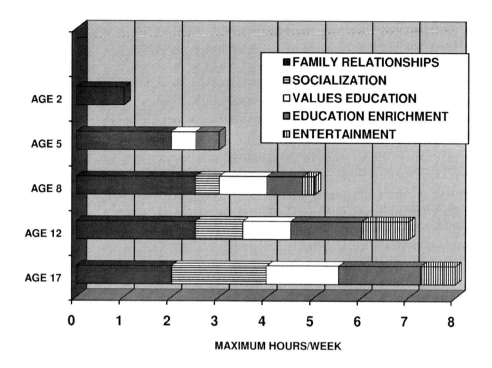

Figure 8.1: For ages 2 through 17, here are guidelines for weekly hours of consumption of interactive media.

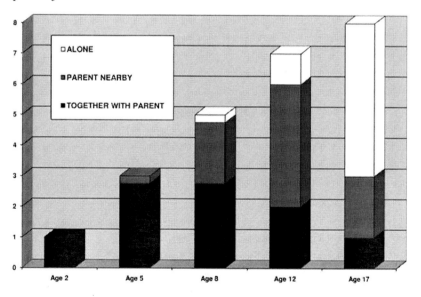

Figure 8.2: For ages 2 through 17, here are guidelines for weekly hours of parent presence while the child is interacting with media.

Chapter 9: Safety and Other Challenges

Interactive media can gift us with richness and beauty and can slime us with oozing sleaze. Safe interactive media practices at home are not only necessary now, but are also great preparation for a lifetime of interactive media exposure. The good news is that home is where children most frequently use interactive media. Home is the best place to teach rules and correct behavior. Home is where you and your children can solidify and extend Family Relationships.

You are careful about keeping the nursery safe for the baby and you toddler-proof your home. Similarly, you can give thought to being careful about the safety of your kids' interaction with media.

Parents have the home court advantage to teach values and encourage healthy development. When it comes right down to it, short of actually watching older children constantly, parents cannot be absolutely sure about what is happening online, so good basic relationships and trust formed in preschool and the early school years between you and your children are important. Another great tool is your common sense.

Although the methods you will learn later in this chapter do not explicitly address safety, the whole approach is based on your active involvement, encouragement, monitoring, and control of online behavior. Find your own workable ways to monitor and prevent the potential hazards. Become and stay involved. That's mostly what online safety is about. So, keep your children safe and please relax.

It is never too early to learn good habits and to keep up with good safety practices for the family. In this chapter, I start with comprehensive summaries of hazards and safety measures regarding interactive media. Expectant parents can begin getting comfortable with these practices even before the baby comes. For parents with older children, I address discipline, the "savviness gap," addiction, bullying, consumerism, and media awareness. Lastly, I suggest fun hands-on activities to learn about safety as a family.

Older children in the home not raised on the Media Plan from the start may have already formed strong independent habits in using interactive media. It is understandable that these children will offer some resistance to your attempts to gain control over what had been entirely their unchallenged private domain. Although my focus is on babies and young children, I offer suggestions to families with children all ages. Put to use your best parenting practices and settle for a safe, workable compromise. In some cases, you might need to lock away cell phones, MP3 players, flash drives, and other devices. But always follow through on what you say you will do—or don't say it in the first place.

An older brother may find it easier to adjust to a new system if he helps you implement his younger sister's plan. In fact, older children are aware of the hazards of media and commonly seek to protect their younger siblings.

GENERAL SAFETY CHALLENGES

SEARCH WORDS: parent safety kids online Internet

The Internet makes the family computer and other digital devices porous gateways through which almost any type and amount of text, images, and sounds can travel instantaneously and anonymously to and from your home. Yet, most of us still protect our doors and windows better than we do this portal. If the family computer is an appliance like the family car, its safe operation and user and environmental friendliness must be our ongoing concern. Children are vulnerable and naïve, even until early adulthood. Parents often underestimate the limitations of kids' judgment. Just as a car is not a toy, neither is an online computer.

Interactive media can be used via Wi-Fi through airports or even whole cities, or wherever cellular networks are available. Video games can be played online or bought off the shelf or downloaded and played offline. I strongly suggest that you include these potential portals in your planning for safety and apply the same common sense approach of the Media Plan to these as you do to the family computer.

Threat to family life. Next time you go online, especially to unfamiliar sites, notice how often you try to figure out how much trust to put into a site or decide about whether or not to log into the site and share your personal information. Even while on a trusted site, eye-catching animated ads pop up, sometimes declaring you have won something, and usually luring you to connect to another distant and unknown site. Some countries have a site or sites dedicated to reporting Internet and other digital media crimes, including, but not limited, to exploitation of children.

Spam, identity theft, viruses, and other obvious threats are discussed widely and preoccupy safety-conscious parents. However, I believe that much parental unease goes deeper: Parents fear that they are losing their children to hidden, insidious, and powerful forces beyond their reach. They sense that the very fabric of family life may be subtly disintegrating right under their noses in ways they do not quite understand. The poem below attempts to express the nuances of these fears.

A PARENT'S NIGHTMARE: Where Have My Children Gone?

Our cherished kids: We bring such glorious hopes to them.

But today, our youngest son just floats away,
mouth agape, eyes glassy, fingers clicking,
beyond reach, nowhere, everywhere,
beyond place, beyond time, beyond person,
without thought, enslaved, hypnotized, oblivious, dissolved into disembodied text,
sucked into swirling graphic cyberspace,
cuddled by shiny robots grinning somewhere in Eastern
Europe.

And tonight, our only daughter silently slips past shaded windows,
gliding in and out of gloomy cyber-manholes.
Wads of plastic cash clutched in ghostly hands, she tugs
and pleads again, "Please, mom, please let me
saunter to smutty, harsh-lit stores, let me
consume from glamorous rancid heaps of spam."
"Please, please, dad, please let me run Teflon-free
in grimy, gaudy alleys, swallowed whole
by wonderful swarms of generous, voracious pickpockets.
Please, please, pretty please, please let me meet, oh,
lullabying me so gently, swaying, oh,
so hot in city shadows, bejeweled lewd pimps."

Our cherished kids: Like exotic, far-fetched ocean fish,
air-hungered gasping, wide-eyed blinking,
ensnared in wide-cast careless nets.

PARENTING BASICS

INTERACTIVE MEDIA HAZARDS

HAZARDS TO FAMILIES AND KIDS

SEARCH WORDS: Internet crime statistics study report

- Savviness gap—Parents can be ineffective due to uninformed, inconsistent, or indecisive leadership. Teaching and guiding is difficult when students know more than teachers. In addition to narrowing this savviness gap, parents must have a systematic, rational approach to all interactive media.
- Conflict—Some families are troubled or have members who do not cooperate or solve problems well together. Interactive media can become lightning rods for friction in such families.
- Family Violence—Family violence sparked by conflict or inappropriate punishment is by far more damaging than online violence. Family violence can lead to violent children.
- Values Dilution—Family values are diluted by an alluring and sometimes vulgar popular culture, exposure to violence, and poor etiquette.
- Immaturity—A child's or parent's immaturity can lead to unpredictable interactive media use. Children regard the family computer differently than grown-ups, as they do with most everything else they interact with—somewhere between a toy and a grown-up's tool.
- Marketing—see Kathryn Montgomery's *Generation Digital* (M.I.T. Press, 2007) for extensive coverage of this topic. Marketers go over parents' heads to gather information and entice children to want a product.
- Possible Health Risks—Obesity can result from the response to the merchandising of junk food, snacking mindlessly in front of the TV. There have been some questions about brain tumors caused from cell phone use.
- Interactive media Overuse Syndrome (Addiction)—An addicted family member can become isolated from the rest of the family and a focus for conflict. Children should have a range of social and family after-school activities. They should have undistracted opportunities to do their homework. The alluring Net can disrupt a proper balance of activities. A small minority can become seriously impaired with this addiction-like syndrome.
- Poor Social Skills—Poor social skills result from kids' unintentional rudeness and poor etiquette when interactive media behavior is not properly supervised.

- Accidents—Accident rates, already greater among teenagers because of poorer judgment and impulse control, can increase further when they are distracted by interactive media such as cell phones or texting while driving or walking on busy streets.
- Distraction—Distraction from information overload in the form of intrusive, unnecessary cell phone calls, e-mails, and texting.
- Violence—Violence and alcohol use can increase from media exposure.
- Bullying—Children who punish, demean, or shame other children can do so more easily or even anonymously online. The bullied suffer, while the bully goes unchallenged.
- Inappropriate Activities—Viewing pornography, sexual overstimulation, responding to alluring face-to-face contacts by pedophiles, gambling, and even unauthorized shopping are hazards to impulsive and naïve children.
- Misinformation—Unsubstantiated claims, stories, blogs, and testimonials, some compelling and attractively presented, are easily accessible to children. Children are especially naïve at differentiating fact from fantasy or opinion from fact and exercising good judgment about what constitutes professional integrity and good journalism. Even grown-ups may believe inaccurate medical information that may mislead them in giving poor medical care to their child; they may even trust what see online more than their doctor's advice.
- Terrorism—Terrorist groups are increasingly turning to the Internet to recruit youngsters. Cyberspace has become a place where governments fight terrorism. Hate groups are increasingly sophisticated in using YouTube and social networking sites.
- Unknown Long Term Effects—Child development specialists do not yet know how interactive media will affect our children and future generations. For instance, Internet and other media violence can increase children's aggression as well as emotional and psychological problems. If TV is any guide, there can be substantial effects for both the good and bad.
- Prescription Drug Abuse—Prescription drug abuse is a growing problem as more sites and e-mail solicitations offer unfettered access to prescription medications without a doctor's authentic Rx.

HAZARDS TO THE ENVIRONMENT

- Emissions—The IT industry energy consumption accounts for about the same of the world's CO_2 emissions as the aviation industry.
- Paper Waste—Paper is wasted by unnecessary printing.
- Toxins—Discarded computers, ink cartridges, and laptop batteries release of mercury, lead, and other toxins into the environment.
- Energy Footprint – Electricity consumption of electronics, on and off.

PARENTING BASICS

INTERACTIVE MEDIA SAFETY

SEARCH WORDS: safety children online computer Internet software site

Stay involved and informed, foster trust, teach common sense, and practice prevention.

CARING FOR CHILDREN AND FAMILY LIFE

- Follow sound child development principles, including being fully present regularly, being consistent, and following through.
- Lead your family with the Media Plan. Keep in mind that your goal is to promote your children's development and enhance your family life.
- At the very least, restrict the time your child spends with interactive media alone until he shows responsible online behavior.
- Do not place interactive media in the child's room.
- Teach and practice etiquette, kindness, consideration, honesty, and non-violence.
- Join safe virtual communities with your children and encourage sharing information about safety matters.
- Forbid or severely limit violent games.
- Forbid cell phone use on the street to prevent accidents.
- Teach good judgment regarding how much to share online and shutting off media when it is time for homework.
- Teach discipline in the use of the Media Plan.
- Prevent excessive exposure to violence, inappropriate sexuality, as well as drug and alcohol use.
- Block routine Internet access via mobile devices.
- Block or filter sites with inappropriate unwanted content.
- Recognize her abilities and limits so that you can provide an online and interactive digital plan that meets her developmental needs.
- Spell out very specifically rewards and punishments for online and other interactive media behaviors. Be consistent and fair.
- Make sure your children know you want to hear their problems and how important their safety is to you.
- Become computer and media savvy and keep current with technological

changes—many books, magazines, newspapers, and online sites provide such information.
- Do not allow snacking in front of the computer.
- Do not ever allow interactive digital activities to interfere with meals, bedtimes, and other family routines.
- Turn off the TV at mealtimes.
- Keep a regular dinnertime discussion going about the benefits and perils of the Internet and other interactive media.
- Advocate media literacy programs in your school and community for fellow parents and children.
- Learn and teach about the technology of the Internet and other media.
- Immunize preschoolers and shield others from excessive influences of media merchandising by sitting with them and explaining what is going on.
- Do your very best to minimize children's exposure to online and other media marketing efforts and teach them to recognize and avoid them. Support research efforts that will give us better knowledge for the long term.
- Visit sites provided by child advocates for reviews.
- Utilize GPS devices at shopping malls, airports, and when traveling, to keep track of children.

CARING FOR THE ENVIRONMENT

- Emissions—advocate for IT industry energy conservation policies. Use energy-efficient media devices at home, and adjust power management controls to save energy.
- Paper—Use recycled paper, print on both sides if possible, and avoid unnecessary printing. Recycle paper. Reuse printer paper whenever possible and install software that saves paper.
- Toxins—Recycle discarded computer hardware, ink cartridges, and laptop batteries to spare landfill release of mercury, lead, and other toxins.
- Energy Footprint—Turn off and unplug media not in use.
- Have your children join you in these practices to convey to them the value of protecting our resources.

CARE IN ONLINE PRACTICES

- Discuss privacy, ethical behavior, and perils like bullying and fraud, with your kids.
- Forbid dishonest, rude, unkind, and bullying behaviors.
- Teach kids about how advertising works, how to use reliable sources, and identify misinformation. Alert kids about the perils of popular media and

how to become careful consumers of information. Alert children to the many propaganda sites, including terrorist groups that recruit children.
- Teach and re-teach children safe online and other interactive media practices as you would in a manner similar to traffic safety, prevention of alcohol and drug use, and other high risk behaviors. Teach them to identify sites rated by safety-rating services.
- Strictly forbid face-to-face meetings with any stranger from an online encounter, no matter how alluring.
- Avoid downloading anything until you are computer savvy enough to determine its safety. Strictly forbid children to download anything if you are not present.
- Teach children the ethics of downloading music and other copyrighted material legally and avoiding pirating.
- Be present when your child roams the Internet or uses other interactive media, or join him.
- Avoid sites and other media heavy with ads.
- Treat your e-mail address as you do your home address.
- Investigate the truthfulness and security of all commercial transactions. Forbid any interaction which requires revealing private, personal, or family information. Do not give children your credit card number.
- Beware of any e-mail offer—even an offer to cancel unwanted e-mail. Beware of any request for an e-mail address or personal information. Instead of giving your main e-mail address, give another you might open specifically for these purposes.
- Government and the Internet security industry try to stay ahead of ever-growing cyber-crime. Keep up with the latest information, or at least install self-updating security software. For example, the Federal Trade Commission provides extensive consumer protection information. *Consumer Reports* runs frequent articles. Microsoft has a helpful site, as does the American Academy of Pediatrics. Other online resources include help you stay informed about security. The FBI accepts complaints about Internet crime.
- Keep up-to-date with anti-virus and other security measures.

The "savviness gap" between parents and children can be hazardous. If you start the Media Plan early, you can learn along with your child as he grows older. Otherwise, try to narrow this gap. You would not place a potentially dangerous appliance in your home without knowing how it works and how it could be misused. Older children can misuse interactive media in ways they know their parents would not approve. Even a parent in the same room may not have any idea of what the child is doing. Such isolation is not good for children.

To benefit fully from this book, I suggest you have basic skills to operate interactive media devices your child and his friends might use. Keep up with children's ever-changing lingo. Like pig Latin of old or speaking in code, this lexicon can conceal their messages.

Ask your school-age child to teach you his newly-learned interactive media skills. You will empower him and begin an active and positive conversation about his online and media experience that will set a helpful tone and form a good basis of changes to come. Ask him to show you his video games, and play with your child. Listen to his MP3 player selections with him. Have your child show you his mobile phone. Most children will see your efforts as positive and enthusiastically comply.

INTERACTIVE MEDIA OVERUSE SYNDROME or "Addiction"

SEARCH WORDS: kids children Internet addiction computer

The label "addiction" is often used to describe such excessive and driven use of interactive media that it interferes with family, social life, and productivity. These activities can include too much time with video games and online activities like excessive and compulsive roaming, search engine and database searching, shopping, gambling, gaming, involvement with online relationships, and in older kids, sexual behaviors. If such activities persist for more than several weeks and impair proper functioning, the child or teenager may be said to be suffering from an addiction, similar to other behavioral addictions like compulsive gambling, sex, exercise, overeating, and shopping.

Whether digital technology overuse and behavioral addictions actually share biological mechanisms and should be lumped in with chemical addictions is still controversial among medical experts. At this stage, I believe that the term "Internet addiction" is overused and often misapplied. It provides an easy pseudoscientific label, but it does not necessarily lead to genuine understanding and helpful interventions.

Although the Internet is currently one of the main interaction kids have with technology, what will come later? So, rather than "Internet addiction," I prefer to call this very real cluster of behaviors that can affect as many as ten percent of kids as "Interactive Media Overuse Syndrome."

The syndrome brings a serious degradation in the child's quality of life and distorts his time management and balance of other priorities. Adolescent boys suffer most frequently. They spend undue time and are preoccupied when away from it. The teenager immersed in media may appear busy and focused, often performing several tasks quickly, almost frantically, at the same time. He stops only with great difficulty.

Such children may experience a calmness or a pleasant high when online, and withdrawal symptoms like craving and irritability when suddenly stopped.

In addition, he needs increasing amounts of exposure to achieve the same pleasant mood. Conflict with parents about other priorities and within himself about being out of control may be other important symptoms. There may be physical symptoms as well, such as poor sleep, hygiene, and eating habits; headaches, dry eyes, carpal tunnel syndrome, and backaches. A little-known hazard is the triggering of epileptic seizures by the flickering screen.

Such teenagers are often depressed to start out with and their digital technology overuse may actually have some benefits for them. Studies from all over the world suggest that the syndrome can be associated with (not necessarily causing or caused by) depression, suicidal ideation, and other clinical disorders. Kids with attention deficit hyperactivity disorder and other impulse disorders may also be more vulnerable. How much the syndrome is an expression of an underlying disorder and how much it causes a disorder is uncertain.

Here is what excessive immersion in media can look like: An adolescent spends endless hours at the computer in his room. His life lacks the pattern of daily, weekly, monthly, and yearly rhythms, often tied to natural cycles. He lacks a continuous and stable family schedule to orient and anchor him. He has no set bedtimes and mealtimes. His natural rhythms are chaotic and he overlooks the calls of nature for as long as he can. Holidays, birthdays, anniversaries, social calendar, religious holidays, seasonal sports, and other family and community events are distant blurs.

While he may regularly visit some sites, his attention is undisciplined, as he strays from one ephemeral thrill to the next. He lacks a firm grasp of where he is in time, place, and person. In cyberspace, he is often not quite fully aware of where he is, who he is to others, and who are the people around him. His interactive media experiences are intangible, sterile, empty, unsatisfying, and often barely activate his dulled senses. Without sufficient anchoring in tangible and definite landmarks, he is disoriented, lost, and anxious, becoming increasingly frantic in a search for meaning and a home online. There are few reference points and feeble compasses to guide his values and affirm his sense of who he is and where he belongs.

The young man lacks the human relationships that would ground him with real people in time, and real place. He can never be quite certain if a stranger is trustworthy or even real. While he may text several people at once, he is not fully present in any of these encounters. Too much time in this virtual world disorients him and distorts his development. At times he realizes that he is enslaved by powerful impulses, but he feels helpless and is increasingly afraid to risk the challenges of ordinary life.

Doctors are just beginning to study this syndrome, and it is important not to mislabel or misdiagnose a child. Nevertheless, the problem may not be obvious until it is serious, and parents, doctors, and school personnel should be alert. The syndrome has been studied primarily in adolescents. Shy, isolated, and unpopular teenage boys seem most vulnerable, probably because the anonymity of Internet interactions is more productive and comfortable than face-to-face social connections.

If the symptoms persist for more than several weeks and you cannot impact your child's impairment significantly, I strongly suggest you have your child evaluated by a qualified child and adolescent psychiatrist or another knowledgeable professional expert. Make sure you understand fully what your child is going through and are certain that the plan is reasonable before agreeing to any treatment .The affected teen may easily relapse even after seeming to have been cured.

This book offers methods for parents to try to prevent entirely or lessen the severity of this syndrome. I believe that a steady Media Plan approach from early childhood will immunize children from adverse interactive media effects.

CONSUMERISM AND MEDIA AWARENESS: YOU GET THE COOKIE, THEY GET THE TREAT

SEARCH WORDS: market, children, kids, media, ads

As consumer education and media awareness are becoming increasingly sophisticated merchandisers have captured young consumers by segmenting markets according to developmental abilities. Novel interactive media are now affording marketers more effective means for directly targeting babies and other young children and bypassing parents to form individual relationships that shape their consumer preferences. Parents and children are put in a difficult position, and family ties are weakened as conflict arises when a child stubbornly insists on a product seen advertised on TV.

We owe much of our freedom and success as a society to the benefits of free markets. An essential part of a market is the seller's legitimate attempts to learn about who is going to be his best buyers and then attract and sell to them. Many marketers use ethical and transparent methods to find out about us and to divide us into market segments. The data about us may give them an advantage in intense competition for our attention and purchasing dollar, as they pitch their products differently to different market segments.

However, there are some marketers who use deceptive practices. Interactive media can provide marketers direct access to individual buyers for data collection and direct sales. We live in a wonderful society where we have many choices of goods and services to choose from. Never before in the history of mankind have there been such a plethora of material goods and merchants clamoring to convince us to buy them. And the merchandisers know what they are doing.

Marketing is now an applied science that encompasses sociology, psychology, and anthropology. And the youth market is a multi-billion dollar industry, working its way to younger and younger consumers directly, through parents, and even through school curricula, making it increasingly necessary for parents to teach kids to become wise consumers. In *Buy, Buy Baby* (pp 193-206), Susan Gregory Thomas describes the inroads marketers have made into the daycare industry.

Children are vulnerable. Advertisers use sophisticated technology to detect consumer practices. Just recently, as the mobile phone industry is looking for new markets, it is aiming at younger and younger children. "The youth market is particularly enticing because these customers treat their mobile telephones more like a companion than a device—or like a 'doudou' or stuffed animal," as AFOM, the French mobile telephone operators trade association, described it in a report on customers' habits in a summer survey. "In general, young customers chatter more on the phone, spending more on the latest games, ring tones, and wallpapers." Additionally, more and more children can go online using smartphones and laptops.

For maximum effect, marketers design ads tailored to children's levels of development. They increasingly directly peddle to children and bypass parents, disturbing important family roles and boundaries.

Research on interactive media and children is just beginning, but we can extrapolate from what we know about television's influence in our homes over the last half-century. An excellent readable summary of the impact of TV is *The Elephant in the Living Room* by Drs. Christakis and Zimmerman, child development specialists at the University of Washington. Marketers easily exploit the immaturity of children to influence parents' buying decisions. In some ways, the distinctions among media types are already blurred, because television, online computer, video game console, telephone, music, photo storage, and camera and photo processing are all becoming integrated into single devices.

The Internet provides vast opportunities to marketers: active selection, interaction, communication, and access with much less government regulation than TV. Significantly more families with children have an Internet connection than those without kids. Kids often prefer to go online than watch television. New hardware is adapted for kids. Video games bring the arcade into the home. Children are fascinated by the novelty, reach, power of the technology, and the variety and innovation of the contents. They readily welcome new interactive media and are eager for new versions. Naturally, these facts are appreciated by marketers and merchandisers who want to promote their products. Marketers are eager to use personal information mined from the child to tailor a personal approach to him.

Susan Gregory Thomas describes the extent of marketers' reach into the nursery and playpen in *Buy, Buy Baby*. She describes the rise of marketing/entertainment/manufacturing/media megacomplex that promotes toys and other products directly to kids as young as two, with enormous success, often bypassing or seducing parents. Often, moms of Generation X are also targeted. This megacomplex employs experts in developmental psychology to design their approaches.

While parents can be powerful influences in determining the role of consumerism in their children's lives, it isn't easy. I want to empower parents to take back from marketers the training of their children to be intelligent consumers who are not unduly influenced by the intense, expertly designed merchandising that saturates our lives.

For starters, examine your own attitudes toward advertising and behavior toward brands. Your choice of a brand can be personal, expressing who you are. Some people have intense emotional brand loyalties. Others consume carefully, reading reviews and deciding what would suit their needs on the basis of as much unbiased information as they can. Which are you?

Are you especially loyal to some brands? Do you know how you have formed these loyalties? Do you go out of your way to show off your loyalties by wearing, carrying, or eating a product that announces your brand preferences? Would you want your baby to start following in your footsteps? She surely will if you do nothing but respond to merchandisers uncritically. Since you can affect what your children learn, would you consider changing some of your attitudes and behaviors? A wonderful resource for beginning to help you is the Center for Media Literacy.

LEARNING ABOUT SPAM TOGETHER

Children learn best from hands-on activities and projects with a patient parent who is patient. Here are some experiments you might enjoy together.

Experiment 1:
How safe is random roaming online?

Ask children six and older to help you. Explain each step and answer questions. Allot at least an hour.

1. Explain to your children that you will learn together how safe going online is, what you will do next, and what might happen. Each participant can guess how many intrusions will occur.
2. Before going online, update and run your spyware and adware software, so that you know your computer is wiped clean.
3. Now, visit sites you know and some that are new to you and your kids. Leave all your firewalls and protective software in place.
4. After some time, log off and clean your computer again as you had earlier.
5. You will now discover several unwanted attacks on your privacy—I did.
   My spyware and adware are current and highly rated, yet I still had several.

One cookie came without permission while I downloaded and installed a toolbar from a well-known, trusted site. When I called the site's Web master, he denied his site installed any cookies. Be careful.

Experiment 2:

PART A: How safe is answering unsolicited e-mails?

Ask children six and older to help you. Explain each step and answer questions. Allot at least an hour.

1. Tell your children that you will learn together how safe going online is, what you will do next, and what might happen. Each participant can guess how many intrusions will happen.
2. Explain each step and answer questions.
3. Before going online, update and run your spyware and adware software, so that your computer is wiped clean.
4. Open a new e-mail account at one of the service providers.
5. Elect not to use the site's spam protection if offered, but do use the site's antivirus protection.
6. Sometimes this in itself can be an unsafe step, so update and run your spyware and adware software to see for yourself.
7. Starting with a clean computer, respond to an unsolicited e-mail offer or two or for a free product or service from the new address. Do not reveal personal information other than your new e-mail address, and do not agree to any terms. Cooperate otherwise, and follow the endless screens and surveys attempting to ensnare you into purchasing something.
8. Log off and clean again.
9. You are likely to find many intrusions. I did.
10. Leave the e-mail account intact for Part B.
11. Discuss the implications with your child.

PART B: How much spam will you generate by answering unsolicited e-mails?

Ask children six and older to help you. Explain each step and answer questions.

1. Following Part A above, tell your children what you will do next and what might happen. Each participant can guess how many spammed e-mails will come in one week.
2. Now, leave the new e-mail account alone. Do not delete any incoming e-mails. Just watch it.
3. A week later, count the number of unsolicited e-mails you received.
4. Leave the e-mail intact for PART C.

It is likely that over the ensuing week and beyond you will be spammed intensely. I was. Although I conducted this experiment several months ago, the new account is still receiving up to fifty unsolicited e-mails daily. Why? Even if you had bought nothing in your little foray, the sites' owners made money by generating your profile, attacking you with spy, ad, or mining software, and selling your email address to other businesses, some legitimate and others not.

This is a wonderful opportunity to discuss with your children ways the family's privacy can be compromised, even when he answers simple questions online like "What is your favorite store?" for a free gift or even goes to a site that requests a cookie.

PART C: *How well will the* **e-mail** *provider protect you?* Ask children six and older to help you. Explain each step and answer questions.

Following Part B above, tell your children what you will do next, and what might happen. Each participant can guess how many spammed e-mails will come in one week.

Return to the site where you set up the e-mail account in Part A and use all the protective measures the site offers.

Repeat Part B, steps one to three.

Decide what to do with the new e-mail account.

The results will depend on your e-mail provider's defensive services.

Comment: We learned that common sense and caution are essential, because even the latest protection cannot immunize you perfectly. There is an ongoing war between Us (good guys, honest people, and businesses, and the Internet security industry) and Them (bad guys, hackers, thieves, con artists, and other sleaze). Beware. Don't be sitting ducks.

But also use these experiments to teach your children other important lessons: What is advertising? How does it work? What is it for? Is it wise to buy something just because an ad is alluring? How do advertisers know what ads to make for whom? What makes for honest business? How should thieves be punished? What does the family's religious practice say about all of this? How can we tell a business is honest? How do credit cards work?

EXAMPLES

Sammy learned the easy way. Using the ideas in this chapter against a background of what he learned earlier about seven-year-old Sammy's developmental needs, Dad planned out a mini course on safety with eight-year-old Sammy. The two spent several hours over a period of weeks learning about each of the hazards together and found online resources to help them stay current. They learned about the workings of the computer and other technical matters in the process. Sammy was appointed the family computer safety monitor, and father and son taught other family members together about online hazards and how to avoid them. Father and son continued regular safety reviews once a month.

Comment: Learning about and monitoring safety is an ongoing project. Involving children provides Growth Opportunities in Family Relationships, Values Education, Socialization, and Education Enrichment.

Chapter 10: Fancy Menus and Special Plans

In the *Setup and Quick Start Guide* you began to transform your family's use of media. In this section, you will take the Media Plan to its maximum power.

Parents of newborns have plenty of time to begin their planning. Older kids require a transition into an effective system. Choose what seems most comfortable and suitable to your family. You can always change later. Remember that you are aiming for progress, not perfection. Like cooking, the tasks ahead are based on trial and error. Mistakes are inevitable and can become great opportunities to learn and have fun.

The examples here are vivid, instructional portrayals depicting how parents can use media deliberately and thoughtfully in raising children and running families. Of course, you can change method to create your own, but try to be consistent at first for at least a month or until the practices you put in place become habits in your family life. Be creative and have fun! *See Setup and Quick Start Guide for less intensive plans.*

HOW INVOLVED WILL YOU GET? (continued from page 48)

Casual: Build on the plans described on page 48.

Interview people who know your child and talk to other parents about your project. Decide about your child's developmental needs and discuss your goals with all family members. Take a proactive leadership role.

Move the computer away from the wall, and put several chairs around it to assure good line of sight contact among users, making sure it is both visible and within your earshot at all times.

Create a plan based on your child's age and needs, according to guidelines for each age group, and as listed atop the Weekly Menu form on page 36. Find sites that provide the content you need for your child, using the tables in this book as guides. Review safety, monitor progress, and readjust your plan every two weeks.

Gourmet: Build on the plans described above.
Study this book thoroughly and take notes.
Obtain as many peripherals as necessary, modifying the physical environment to make it a pleasant place for people to interact with each other around media.

On a weekly basis, allocate activities listed in the Media Plan Worksheet into daily menus, completing the Weekly Menu form. Review safety and monitor progress daily. Keep all your lists updated, provide new menus and readjust your plan weekly.

WEEKLY MENU PLANNING

A food menu is the actual plan you make so that each meal has balanced nutrients from each food group. Meals come in all shapes and sizes. Menu design and meal preparation aim for nutrition and timely and attractive presentation. Planning, buying, and preparing the food for meals that contain balanced basic nutrients and serving them in attractive meals is certainly an ongoing challenge and a major part of being a parent.

Similarly, a media menu is the plan you make for each online session to give your child a balance of Growth Opportunities. Consider attractive combinations of Growth Opportunities when creating online and interactive media menus. Like school lunches, the child may also get interactive media experiences at school.

You will best understand how to use the process by looking at the forms. You have already seen how the Media Plan Worksheet on page 44 summarizes the media activities ideally you are going to allow and encourage. Give blank forms to your partner and children; have them enter their ideas, and then collate the results to make the master menu. You only have to do it once in a while.

Now, let us look at the Weekly Menu form. The top one-third provides a summary of guidelines for all age groups and a way for you to customize them for your child. The remainder is a weekly schedule for allocating the times among the Growth Opportunities. The numbers in the guidelines are just approximations—change them to suit your own preferences and your unique child's needs.

Now you will create the daily menus. Basically, you will divide each Growth Opportunity and time with parent in your child's plan among several daily sessions.

Schedule and plan your media sessions several days ahead of time. Include your child and other family members in scheduling and planning. Prepare for each online session as you would prepare a meal or a trip by putting links together in a special directory or folder ahead of each session.

Dr. Eitan D. Schwarz

WEEKLY MENU FOR_____ WEEK OF _____ BY_____

Hours /Week (Digital)	Age 2	Age 5	Age 8	MY CHILD: AGE ___
minimum Family Relationships	1:00	2:00	2:30	
maximum Socialization			:30	
minimum Values Education		:30	1:00	
minimum Education Enrichment		30	:45	
maximum Entertainment			:15	
total maximum	1:00	3:00	5:00	
total minimum			1:00	
maximum per session	:15	:45	1:00	
minimum with parent	ALL	2:45	2:45	
minimum parent nearby		:15	2:00	
maximum independent			:15	
+ Hours /Week Other Media	3:00	6:00	10:00	

www.mydigitalfamily.org

DAILY MENUS

MONDAY ____ hour(s) ____ to ____
w/parent   near: alone:
Family Relations
Socialization
Education Enrich
Entertainment

TUESDAY ____ hour(s) ____ to ____
w/parent   near: alone:
Family Relations
Socialization
Values Education
Education Enrich
Entertainment

WEDNESDAY ____ hour(s) ____ to ____
w/parent   near: alone:
Family Relations
Socialization
Values Education
Education Enrich
Entertainment:

THURSDAY ____ hour(s) ____ to ____
w/parent   near: alone:
Family Relations
Socialization
Values Education
Education Enrich
Entertainment

FRIDAY ____ hour(s) ____ to ____
w/parent   near: alone:
Family Relations
Socialization
Values Education
Education enrich
Entertainment

SATURDAY ____ hour(s) ____ to ____
w/parent   near: alone:
Family Relations
Socialization
Values Education
Education Enrich
Entertainment

SUNDAY ____ hour(s) ____ to ____
w/parent   near: alone:
Family Relations
Socialization
Values Education
Education Enrich
Entertainment

TWO FAMILIES

Here are two families following the more detailed methods outlined in the Casual and Gourmet methods above. First, let's meet them and see how they differ. Then we will see how each approaches the challenge of putting a Media Plan into place in ways that match their lifestyles. While the basic plan for two boys are very similar, their applications suit their family's lifestyles. Then, finally, we will see how the children play together while each adheres to his own plan.

The Smith's home is lively and busy. Mr. Smith is a sales manager, and Mrs. Smith works part-time as an interior decorator. Lisa, their thirteen-year-old going on thirty, has her own room, while Michael, four, and Danny, seven, share bunk beds in their bedroom. An assortment of pets enlivens the Smith home. Each child has differing needs, and obviously Lisa's are more advanced than her younger brothers. Danny is more mature than many seven-year-olds, and Michael is more babyish than other four-year-olds.

The Smiths have limited space and resources. After reading this book, it made more sense to allocate funds for a full-featured online family computer with all the bells and whistles. They decided to place the computer in the family room near the kitchen, configured and arranged for multiple users as a "media center." For more flexibility, they installed a wireless network to connect their new machine with their ancient five-year-old workhorse upstairs. Everyone could still use the old computer for word processing and other utilities.

This was the logic they used for sharing their one car or planning family vacations. Mom and Dad could now use the old workstation with a password to go online via the network for e-mailing, bill paying, online auctions, and other activities. They also sometimes allowed Lisa to use the workhorse for her increasing private online use.

Having three children made the Smiths' job more complex, and neither Mr. nor Mrs. Smith had the temperament to be as meticulous as their neighbors, the Roberts. The Smith household was much more casual, and the atmosphere very different from the subdued mood at the Roberts's. The Smith kids were generally more on their own, and there was more commotion as they interacted and the family pets wandered about. But the Smith children were good kids and knew how to compromise and respect their differences.

Now, let's meet the Roberts. In many ways, they are very different from the Smiths. The Roberts have one child, seven-year-old Sean. Dad is an accountant. Mom was an accounting major in college and carefully uses all the features of home accounting and tax-preparation software. Their home is meticulous inside and out, as are their clothes, and their car, which always shines inside and and out. Their lives are orderly, with their schedules posted on the bulletin board. They plan and research major purchases carefully. Both parents exercise regularly and are very careful about their diet. They shop in an organic food store and use many supplements. They separate recyclables from their other trash vigilantly.

Naturally, their parenting style is consistent with their lifestyle. For instance, Mrs. Roberts carefully monitors what Sean watches on TV, and he has a regulated daily routine. For the most part, they are warm and loving and encourage Sean's independence, but they tend to be somewhat controlling. Sean's temperament is agreeable and matches that of his parents. He is a good student and athlete, but on the quiet side. He is mostly compliant, but can occasionally be rambunctious and disobedient.

Mrs. Smith has thought to herself more than once that her neighbor was too controlling and babied Sean, while Mrs. Roberts thought that that her neighbor was too casual and messy and did not supervise her kids enough. *Vive la difference!* Although very different in lifestyles, the Smiths and Roberts have been good neighbors and friends for eight years.

Sean is close to Danny, and the children play together and attend the same school. Mrs. Roberts and Mrs. Smith carefully read this book and agreed to make better use of the Internet and other media, especially video games.

The two mothers wanted the same goals, and talked a lot about developing media menus for their children. Mrs. Smith was more creative and global in her thinking, while Mrs. Roberts was careful and detail-oriented. They worked well together to become savvy, to scout online resources, and to compare notes all along the way. They even came up with similar items on their Media Plan Worksheets. But when the time came to prepare the actual menus, they parted company and approached their tasks very differently.

A CASUAL METHOD FOR A CASUAL FAMILY

Mrs. Smith liked the idea of selecting sites appropriate for each child. Using Media Plan Worksheets, she sketched out three separate lists of Growth Opportunities, one for each child (Danny's example to follow). Some sites were good for young Michael, and some only for Lisa. Some sites were OK for all.

She also liked the idea that each child would spend different minimum times, total times, and alone times weekly, as well as different proportions of time on each Growth Opportunity. Mom, Dad, and children each signed the form to indicate agreement on the times and sites. Most of all, she liked the idea of making media use purposeful.

She figured out those times for each child using the table in the box of the Weekly Menu form (see example). But she found the task of actually allocating and scheduling overlapping daily menus just too daunting, and Dad wasn't much help either. Besides, that just wasn't their style. So she did not go that far. She stopped at planning approximate weekly times.

Fancy Menus and Special Plans

Mrs. Smith held a family meeting, and together the parents introduced the idea of their new media center as a powerful tool and family appliance. They encouraged the kids to search for new sites they wanted to add to the approved list. They reminded the kids of the usual rules for sharing, and that older kids and younger kids have different needs and privileges.

They introduced the rules for online safety that apply to everyone and the consequences of violating these rules. They had Michael post the rules (see page 72) on the family's bulletin board and at the computer. They made sure that the kids knew that a monitoring system would reveal their roaming history and that filtering software and timers will also be used. Mom spot-checked the logs, and depending on the nature of a violation, would give consequences ranging from losing Entertainment time to more serious consequences.

The Smith parents described how the Media Plan was going to work for each child and emphasized that it was the same as other family rules and expectations. Total time and time per session were to be set and not negotiable, except in very special circumstances. Other times were somewhat negotiable, but only with a special meeting, and not on the fly, because banking or accruing time would be too difficult for the parents to track. Mom said that there were some sites that would suit more than one child, and if they shared this time it would count toward Family Relationship time as well.

The parents emphasized that all these rules would apply to the use of mobile phones or other hand-held media devices and also when visiting another child's home. The parents stated that they would be spot-checking and monitoring what was going on and were specific that a major offense, especially when followed by a cover-up, would result in a long suspension of online privileges and major groundings. The parents showed the kids they were united and in total agreement. Mom then said that she would get together with each child individually to agree on his or her Media Plan details. The kids asked questions, and the parents were patient and helpful. They pointed out that other families might not have this system, and that their family is ahead of many others.

Mom and Dad met with Lisa first. Father mostly sat there, giving his approval and agreeing with Mother. This meeting was a continuation of an ongoing conversation, because Mom had already been empowering Lisa by learning from her about computers and the Internet. Lisa had already helped Mom develop her list on the Media Plan Worksheet on page 44 and had made suggestions for her brothers. Mom adapted guidelines given for kids Lisa's age (see page 36) in making Lisa's Weekly Menu form. Nevertheless, Lisa was a budding teenager and could be quite sassy and stubborn, so Mom approached her with respect and tact.

Mother and daughter again went over Lisa's list to make final changes. Mom told Lisa firmly to adhere to her list of approved sites. Lisa grumbled and rolled her eyes. Mom showed Lisa the box from the Weekly Menu form. She explained the table and her thoughts about the time limits. Lisa was a bit more convinced, but Mom found a line between flexibility and firmness that (mostly) worked. They made last-minute adjustments, and Lisa wrote the amount of time for each Growth Opportunity that she would aim for in a week.

Lisa agreed that she would work online with her brother Michael for thirty minutes a week and that would count toward her Family Relationships time. Mom repeated the rules; the expectations that Lisa would adhere to the times and sites suggested, and the consequences for infractions.

Mom repeated the same format with each child, but made Media Portal Pages (see page 45) for the two younger boys using her word processor. The example of the forms that follow are Danny's. Following my suggestions, she approved an additional eight hours a week of non-digital media time like movies, TV, and sports events. She planned to approve activities that provide balanced Growth Opportunities.

Once the Smiths had met with all three children separately, the family again met together to exchange more ideas and coordinate and revise the menus again. They reviewed schedules to make sure they would work for everybody and that the parents would be available to spend time online with each child. With much fanfare, each child and both parents signed the bottom of the child's Media Plan Worksheet to acknowledge the importance of the new arrangement, and the other children added their signatures as witnesses.

Danny placed several chairs around the family media center. From there on, the Smiths followed the same procedures as the Roberts next door, except that the Smith kids occasionally tested their parents' patience, convictions, and rules. On the whole, the Media Plan fit well into the Smith household.

A great Saturday afternoon. At lunch one Saturday, Danny asked his father how corned beef is made. The two hurried home and researched the subject on an online encyclopedia. As they followed the links provided in the article and downstream sites, they talked and joked with each other. Before they knew it, they had spent two great hours together. His parents were pleased that Danny benefitted from additional Family Relationships and Education Enrichment Growth Opportunities.

Later, Danny showed Sean some of the sites he had visited earlier with his dad. Since the Roberts kept Sean on a short leash, he told his parents that he had gone online with Danny and described what they had done. Sean knew his parents were fair, but he also knew that they strictly adhere to schedules and family plans. So they had a conference. They weighed several factors: Saturday was supposed to be Sean's day offline. So Sean's parents decided not to count Sean's online time toward his weekly allotment. Sean just got a free hour. The whole process enhanced the Roberts's family life as they made the decision together. Sean saw how fair decision-making requires careful weighing of circumstances.

MEDIA PLAN WORKSHEET For *Danny*

TOTAL ONLINE TIME / WEEK 4:15 HRS **MAX** :45 HR / SESSION

ALONE :15 hr **PARENT NEARBY** 2 hr **WITH PARENT** 2 hr

FAMILY RELATIONSHIPS *CAN BE WITH MICHAEL,*

SOCIALIZATION texting

VALUES EDUCATION The Red Cross, Ourchurch.com

EDUCATION ENRICHMENT www.whitehouse.gov, Virtual tour of Alaska—grandpa would like it

ENTERTAINMENT Video game

NON-DIGITAL MEDIA Movies, TV

We pledge as a family to always show respect, kindness, patience and to use the computer safely. We all agree not to snack in front of the computer and not to let online activities interfere with bedtime, mealtime, or other times. We pledge to help each other keep this schedule.

ALL REVIEWED AND AGREED TO SAFETY RULES

Signed   Danny   **DAD** *Lisa* **Mom**   Date 6/7/07

www.mydigitalfamily.org

Dr. Eitan D. Schwarz

WEEKLY MENU FOR   Danny   WEEK OF   March 5   BY   Mom

Hours /Week (Digital)	Age 2	Age 5	Age 8	**MY CHILD: AGE 7**
minimum Family Relationships	1:00	2:00	2:30	**2:00**
maximum Socialization			:30	**:15**
minimum Values Education		:30	1:00	**1:00**
minimum Education Enrichment		:30	:45	**:45**
maximum Entertainment			:15	**:15**
total maximum	1:00	3:00	5:00	**4:15**
total minimum			1:00	**1:00**
maximum per session	:15	:45	1:00	**:45**
minimum with parent	1:00	2:45	:45	**:45**
minimum parent nearby		:15	2:00	**2:00**
maximum independent			:15	**:15**
+ Hours /Week Other Media	3:00	6:00	10:00	**8:00**

REVIEW SAFETY

www.mydigitalfamily.org

A GOURMET METHOD FOR A METICULOUS FAMILY

Unlike the Smiths, and most other readers, the Roberts worked on a more detailed plan. The Roberts wanted to use the most careful precise approach to media. They worked hard to reform Sean's online experience according to this book and customize a plan that best to fit their son's developmental needs. Here is how they did it.

Mom, Dad, and Sean brainstormed about ways to make the physical arrangement more suitable to interpersonal face-to-face interactions. They purchased a larger monitor; moved the computer away from the wall; and placed chairs in a semicircle to allow people to sit around the small table. From the kitchen, Mom would be within earshot and could keep an eye on the area.

They identified and previewed sites and tabulated them by Growth Opportunities on the Media Plan Worksheet on page 44 along with weekly times. Sean, Mom, and Dad signed off on these as the only approved sites and age-based guidelines, and added eight hours a week for TV and other non-digital media times, listing the shows, sports events, and movies for the coming week and when one of them will join Sean. Dad will spend thirty minutes reviewing safety practices and the computer's security measures with Sean.

Mom worked with the Weekly Menu (see example). She used the age-based guidelines for balanced plans to determine Sean's daily menus. She checked out her plans with Dad and reviewed the forms to Sean to get his ideas and to include him.

Since Sean is seven, Mom used times between those found in the columns for ages five and eight. Mr. Smith and Mrs. Smith also compared notes and agreed on the basic allocations. She allocated extra time for sleepovers to encourage Sean's Socialization. But she also felt that Sean was mature for his age and could have more independent and total time online.

Since Sean was particularly interested in current events and spent much time on news shows and documentaries, she added more maximum and online times, providing opportunities for Family Relationships, Values Education, and Education Enrichment. In fact, she decided that she would even allow more time on this site whenever Sean really got into it.

Mom allocated Sean's alone, parent nearby, and with parent times. For Friday night, she scheduled visits to sites providing Entertainment and Socialization opportunities for Sean and Danny's sleepover. She scheduled time for Sean and his dad and grandfather. She made sure to schedule herself and Dad both to work with Sean and to remain nearby to supervise the boys. Mom added movie and TV football times at the last moment.

Mother worked over the next few weeks to fine-tune and to accustom the family to Sean's new menu. Sean liked being asked for his opinions, especially since his preferences were implemented. However, he resisted the limits on Entertainment and Socialization times, and his parents agreed to extend these as rewards for good school performance.

Dad and Mom used software to restrict Sean to approved sites and timers to limit his total online time. They gave Sean permission to explore the Internet when he studied and spent time with him reviewing the new sites he discovered. In this way they were able to teach Sean their preferences and values in making online choices.

Father reviewed the log generated by software with Sean to make sure that Sean had not wandered into uncharted territories or overstayed at certain sites. Rarely, Sean would cheat and violate the family deal. His parents worked out a system that they had discussed with Sean prior to implementation whereby he would lose Entertainment time if he violated their agreements. Dad and Sean also checked out, updated, and upgraded the protection systems on their computer together. As time went on, they tweaked the physical arrangement to make the equipment and space even more suitable to their needs.

Fancy Menus and Special Plans

WEEKLY MENU FOR_Sean WEEK OF Jul 9 BY Mom

Hours /Week (Digital)	Age 2	Age 5	Age 8	**MY CHILD:** **AGE 7**
minimum Family Relationships	1:00	2:00	2:30	2:30
maximum Socialization			:30	:15 except sleepovers
minimum Values Education		:30	1:00	1:00
minimum Education Enrichment		:30	:45	1:45
maximum Entertainment			:15	:15
total maximum	1:00	3:00	5:00	6:15
total minimum			1:00	1:00
maximum per session	:15	:45	1:00	:45
minimum with parent	1:00	2:45	2:45	2:45
minimum parent nearby		:15	2:00	2:00
maximum independent			:15	1:00
+ _Hours /Week All Media_				8:00
REVIEW SAFETY				:30

www.mydigitalfamily.org

MENU

MONDAY **1** hour(s) **6** to **7**
w/parent   **1**   prnt near: alone **0**
Family Relations :30 **cnn.com w mom**
Socialization
Values Education :30 **cnn.com**
Education enrich
Entertainment

TUESDAY **1** hour(s) **6** to **7**
w/parent   prnt near :30 alone :30
Family Relations
Socialization :05
Values Education :20 **CHURCH.org**
Education enrich :30 **nature.com - mom**
Entertainment

WEDNESDAY **1:30** hour(s) **6** to **8:00**
w/ parent **1:30**   prnt near: alone
Family Relations :30 **cnn.com w mom**
Socialization
Values Education
Education enrich
Entertainment **1 watch TV w mom**

THURSDAY **1** hour(s) **6** to **7**
w/ parent   prnt near   :30
alone :30
Family Relations
Socialization :05 **texting**
Values Education :10 **discuss bully**
Education enrich :15 **nature.com** :15 **mapquest**
Entertainment

FRIDAY **1:00** hour(s) **8** to **9:30**
w/ parent   prnt near **1:30**   alone
Family Relations
Socialization **:45 IM'ing**
Values Education
Education enrich
Entertainment **:45 games.com or other**

SATURDAY **3** hour(s) **2** to **5:30**
w/ parent **3**   prnt near:
alone **DAD**
Family Relations **3:00 TV FOOTBALL**
w/ all
Socialization
Values Education

Education enrich :*30 review security measures w Dad*
Entertainment

SUNDAY **4:30 min** hour(s)
w/ parent**4: 30** w/prnt **4:30** near alone **grandfather**
Family Relations **4:30 Alaska virtual tour / MOVIE**
Socialization
Values Education
Education enrich
Entertainment

NO PAIN NO GAIN: THE HOW-TOS OF DAILY MENU PLANNING

Like the Roberts, some parents will use gourmet method and continue to complete a full week of daily menus in the Weekly Menu form. So how did the Roberts' manage the task of dividing the weekly totals into the daily menus? Let us take a look. Certainly, this method is a bit cumbersome, and not for everybody. Hopefully, it will soon be automated with software to walk parents through the steps easily (see Chapter 14).

CREATING GOURMET MENU

ALLOCATE TOTAL AND PARENT TIMES

Divide the **total maximum** hours/week (Digital) among the days of the week - in Sean's case 5:15 hours.
Remember not to exceed the maximum time per session.
Remember to schedule sessions that require parental presence at your convenience so that you may sit undisturbed with the child.
Allocate your child's time with you, with you nearby, and time alone.

Hours/Week (Digital)   Age 8   Age 13   MY CHILD: AGE 10

MONDAY 1 hour(s) 6 to 7
w /parent :**30** prnt near :**30** alone
TUESDAY 1 hour(s) 6 to 7
w/ parent   prnt near: **1**   alone :

Fancy Menus and Special Plans

```
WEDNESDAY 1 hour(s) 6 to 7
w/ parent    :30      prnt near:    alone  :30
THURSDAY 1 hour(s) 6 to 7
w/ parent       prnt near :30 alone  :30
FRIDAY 1 hour(s) 8 to 9:00
w/ parent  1 prnt near   alone
SATURDAY   hour(s)    to
w/ parent       prnt near:     alone
SUNDAY 1:15 hour(s)  3 to 4:15
w/ parent 1:15 prnt near:     alone
```

Check to see that the times add up. For example, if the total time for a session is one hour, make sure that 'with parent', 'parent nearby', and 'alone' times add up to one hour for that session. Additionally, all daily times with parent, parent nearby, and alone should add up to the times in your child's column inside the Weekly Menu form. This may take some juggling.

ALLOCATE GROWTH OPPORTUNITIES

Now ration these one by one to days of the week by scheduling a site into the daily listings of Growth Opportunities in the Weekly Menu form. Continue to distribute the sites among several days.

Hours/Week (Digital)	Age 8	Age 13	MY CHILD: AGE 10
minimum Family Relations	2:30	2:30	*2:30*

```
MONDAY 1 hour(s) 6 to 7
Family Relations :30 cnn.com w/ mom
TUESDAY 1 hour(s)  6 to 7
Education enrich :30 nature whitehouse.com w/mom
WEDNESDAY 1 hour(s) 6  to 7
Family Relations :30 cnn.com w/ mom
SUNDAY 1 hour(s)
```

Family Relations 1:00 Alaska virtual tour w/ grandfather

Remember, a site can be placed in more than one group. If the total weekly time allotted for Education Enrichment is forty-five minutes, you can assign CNN.com to one session, or divide it among more than one session, say two sessions with fifteen minutes for Monday and thirty minutes for Friday, or three sessions, each fifteen minutes. Or you may schedule CNN.com as offering one or more Growth Opportunities on one or more days. Education Enrichment may overlap with online homework time and instant messaging with collaboration on schoolwork. Be flexible in working with the child in determining exceptions and, whenever possible, empower your child to learn how to make these judgments herself.

Check to make sure that all the times add up. For example, if the total time for a session is one hour, the times spent in all the sites providing all Growth Opportunities that session should add up to one hour. Additionally, all daily times of Growth Opportunities should add up to the times listed in your child's column inside the Family Plan box.

MONDAY 1 hour(s) 6 to 7
Family Relations :30 cnn.com w/ mom
Socialization
Values Education :30 cnn.com
Education Enrich
Entertainment
  w/ parent *:30* **prnt near:**   **alone :30**

ADD OTHER MEDIA

See Sean's final Weekly Menu for how his parents added eight hours of non-digital media: An additional TV hour with Mom Wednesday, a three hour TV football game with Dad Saturday, and a three hour movie with Grandfather Sunday.

NOTES

- Give your child some flexibility in accumulating times for later use.
- Use software to time, limit, monitor, and log your child's online excursions and review the logs together with your child regularly.
- Be flexible and encourage innovation. For example, add to the approved list sites your child has discovered. Encourage curiosity, intellectual pursuit, and exploration.
- On the other hand, if the child shows a pattern of poor judgment and repeatedly tests limits, treat him more firmly, treat him more firmly with a tighter set of consequences, ranging from deducting Entertainment times to groundings.
- This is such a good opportunity to empower the child to learn good judgment about the risks and rewards of using media.

Part 3: FOOD FOR THOUGHT

In this section, I examine in greater depth aspects of the impact of technology on children and their families. Technological products have built-in obsolescence. Not so with your child, or this book: Human nature does not change, and we must deal with how interactive technologies will continue to develop and reach deeper and deeper into the cradle.

Therefore, beginning how and why children naturally play, I look at how technology's major pitfalls for certain age groups. II extend this discussion to novel applications in children's play therapy. I then indulge in wishful thinking about the future of child- and family-centered commercial uses of interactive technologies, automating the creation of Media Plan. Finally, I offer a script for a video game that promotes healthy child development with the hope of showing industry that it is possible.

I devote space in the following chapters in proportion to the importance of the future. As far as interactive media goes, there is a much larger future ahead of us than a present around us and a past behind us.

Chapter 11: Technology and Play: Not the Same Game

Chapter 12: The Future I: Play Therapy and Other Applications

Chapter 13: The Future II: FITGOALS™

Chapter 14: The Future III: The Good Life—A Nonviolent Video game

Chapter 11: Technology and Play: A Different Game

A CLINICIAN'S VIEWPOINT

My goal is to provide parents with practical ways to turn young children's media lives into Growth Opportunities and family assets. This need has become apparent over the past several years as computer, smartphone, Internet and other media applications increasingly mediate everyday human interactions. There is a lack of practical, development-based efforts to assist parents in integrating media into family life. Many more tools are needed, especially to serve the needs of all families and children beyond the middle and upper classes described here.

I am a clinician with over forty years of face-to-face "flying time" with patients—kids, parents, families, schools, and social service agencies (about 70,000 hours) in the science and art of psychiatry. I keep abreast of credible new knowledge and am an expert at making tough decisions covering the spectrum of mental health issues in the real world. Like all other clinicians, I often work with only partial knowledge.

It is inevitable that more of us are now studying the powerful genie of technology, that is out of the bottle and roaming in the lives of families. The events that are taking place under our very noses right now, as media grows explosively and intersects with and child development it, will surely shape the course of our culture and history. But research takes time, and parents need guidance now.

I am well aware that the intersection of children's development, media, technology, and marketing is fraught with controversy. Almost any position voiced in turbulent area would inevitably become distorted and polarized, eliciting lively debate, if not vicious criticism. But there is so much at stake. As billions and billions of dollars in products aimed at children as young as six months flood the market, scientists, child advocates, and many parents are uncertain about the kind of people we are raising and the future of our culture.

As a clinician, I realize that I offer this book on the basis of partial knowledge. We know a lot about human behavior and what humans need to become productive members of families and societies. We know less about early development and the human brain, but we are learning every day as these areas receive more attention from good scientists.

Although our knowledge is as yet incomplete, we must start now to apply what we know to improve the lives of families and children immersed in a sea of technology. The bedrock issue is: How does saturation of youth with media affect development and human relationships in the family and beyond, and how can these wonderful technologies be put to use to promote, rather than harm them?

TO MY COLLEAGUES AND OTHER PROFESSIONALS

More and more educators, developmental psychologists, psychiatrists, pediatricians, and other doctors and scientists now believe that it is our responsibility to become better educated about how media affects children of different ages, to routinely inquire about the quantity and quality of the exposure to media of children in our care, and to regularly recommend healthy practices. In a recent e-mail discussion among child psychiatry and other child-care professionals, several colleagues noted the increasing urgency and frequency of these questions and discussed their responses.

They continue the general restrictive approach. For difficult cases, they suggest abstinence, increasing alternative activities, and other quick fixes. None seems to view children's media activities in the context of play as opportunities for development.

A main goal of this book is to alert child-care professionals to monitor closely and intensively how interactive media impact children and families and help them use these tools in their work.

I believe it is time for us to enter the debate and offer our thoughtful voices. I also aim to alert the IT industry, thought leaders, and policymakers to consider these probabilities in their own long-term planning, prioritization, investment, and innovation. I hope that readers of this book can begin thoughtful discussions that will benefit all. To promote the collegial thinking about interactive media in our clinical work, I describe the methodology, its theoretical basis, and its responsible application, providing examples from my own work.

PLAY

Remember playing as a very young child? Remember how intensely engrossed and excited you felt? Remember how you felt in grade school, especially during recess or visits with playmates? Your craving for play? How serious play was to you? How urgently you felt the need to play? Do you remember how playing was what your life was largely about and the way you connected to almost everyone? That learning happened best when you experienced it like playing? How immersed you became in a special place in your mind, pure and clear, neither quite real, yet not unreal? Well, your child is feeling all of this now. Don't forget what it is like.

How about now? Doesn't being engrossed in a movie, a book, or a ball game feel the same as playing used to? Like a child playing, you are not focused on all reality. You give no thought to the complex business behind a movie; that it is a commercial illusion product, manufactured by hundreds of artists and technicians, shown through a projector on a screen.

Without partially suspending what you know is real, without entering that special space in your mind, you cannot fully enjoy a film, book, or sport. Knowing that it is not really "real," yet suspending that knowledge is part of play. Multiplied by a thousand, that is what play is for your children, starting in infancy. A child playing is a child conducting thought and action experiments, rehearsing for life, and coping. From infancy onward, playing is the work of children. Remember how important it is.

Play activity and materials depend on where the child is developmentally. Regardless of whether the task is education, taking a bath, or eating, the younger the child, the more she regards it as play. The younger the child, the blurrier the distinction between action and thought, reality and fantasy, words and gestures, play and work.

Very young children cannot tell the difference between a toy and a "not toy" because this ability develops in the brain, along with other cognitive capacities, by about two to four years of age. It is about that children realize that toys belong to a special class of objects set aside exclusively for them use in their play activities; toys differ from other things, defined by their parents as "not toys" that belong to the often forbidden and mysterious world of grown-ups.

Older children usually play with toys and do not play with "not toys." Any object can be used for play—a stick, piece of paper, or an elaborate computer game. A spoon is something to clutch or mouth to an infant, but to an older child it is a means to play at feeding her doll, while at the same time an eating utensil and a "not toy" to her mother.

When taking a walk in the park and pretending to be adventurers, the park itself is a kind of play material. When arm wrestling, the body and its strength are play materials.

Older children usually select a 'pretend' toy replica resembling a real object in some ways, but definitely not that object itself, as a prop to facilitate a wide range of play, like dolls and action figures and the houses and cars they use.

Fairy tales and fables are a form of intellectual and emotional play. Safely tucked in a parent's lap or in bed, a youngster readily places himself deep inside the story, eager to experience the full range of what the characters experience. It is as if he reinvents the narrative for himself and uses it to understand himself and the world.

Still older kids increasingly emulate more adult activities, but are actually playing in the sense that they are experimenting with new skills. Older children recognize toys as props meant exclusively for play. The social status conferred by a toy is often part of its value for older children who are learning how social groups and rites of passage work. For them, the closer a toy resembles what adults use, the greater the social status, as they strive toward being more grown up. The use of any specific play material or activity can usually be regarded as appropriate as long as it is helpful, but it should always be safe and sufficiently sturdy to stay intact while the child plays.

Even the adolescent's increasingly autonomous social, sexual, intellectual, moral, and aesthetic decisions continue to be regarded as experiments, as play. Having trendy clothing, hairstyle, makeup, or tattoos and are other instances of play in the service of mastery of the complexities of Socialization and sexuality.

Civilized societies recognize that youngsters spend most of their time playing, so they are not held fully or legally accountable. During this childhood moratorium, children obviously do not receive the same consequences as do identical actions of grownups.

To older children, as to most grownups, the social status conferred by an object is part of its "play" attributes and is not necessarily connected with its actual purpose. For example, driving a car is often more play than convenience. Does a BMW get you to the store any better than a Ford?

TECHNOLOGY AND PLAY

Dolls and their brother action figures have been children's universal companions and playmates for centuries. Children rehearse and explore life's scenarios and challenges and work out ideas and feelings about the world with their dolls. Now, technology has brought a new kind of doll play to children (and adults) in the form of video game avatars. Within the constraints of a game, children can decorate, animate, and motivate their digital dolls to perform ordinary and magical feats and interact with other children across the world. A new class of toys—humanoid robots—will soon populate the toy bin.

Technology has also popularized an intermediate class of toy-like objects among increasingly younger children. For example, in the past, children treasured watches both to learn about time and to gain social status as being like grown-ups. While young children using a wooden toy phone are clearly playing, their older siblings use the cell phone as a hybrid, intermediate between a phone and a toy. Young cell phone users select these hybrid media devices just like they select toys, eager for their non-essential features, like the social status they confer, color, type, and bells and whistles, rather than communication devices that should sound clear and have good battery life. In a sense, they play at using the cell phone and imitating grown-ups. While young children play with miniature toy cars, the teenager uses the automobile as a hybrid. Older children are already eager consumers of new interactive devices and are likely to adopt new ones as they are introduced.

We try to make sure that a youngster has achieved sufficient development to enable the judgment and impulse control to appreciate the "not toy" aspects of the car. The online computer is also a hybrid between toy and "not toy." The younger the child, the more toy-like he regards it, and the more he needs parental involvement to maximize its safe, beneficial use. We must also monitor how much of the "not toy" aspect of the Internet a child can understand.

Today, more than at any other time, older children choose from a wide range of opportunities for interpersonal contact, including face-to-face, video teleconferencing via webcam, telephone, computer or mobile telephone text communication, chat room postings and blogs, e-mails, greeting cards, and actual letters. Older children have found the online computer convenient for meeting people in a variety of ways. Many prefer texting to meeting face-to-face, leaving a message to speaking, or e-mailing to calling.

Like never before, today's kids can choose the modality most suitable to their needs. Each mode of communication offers features that match the child's particular comfort level. As one develops social skills, she experiments with relationships through a greater variety of channels than ever available before. Fundamentally, in the hands of youngsters, these experiments in communication are a form of play. What impact these choices have on development, is yet to be determined.

BABY INVENTS A TOY

Already, smart toys are being marketed that simulate a mother's presence with a recording of her voice to soothe Baby. It is only a matter of time before very young children's play will become a busy marketplace sophisticated, interactive media devices and intelligent toys. We should carefully consider how these may impact our children. Let us begin giving careful thought now about what might work and what may hurt by reviewing our understanding of the brains and minds of very young children.

Healthy brain maturation and psychological development through childhood, adolescence and beyond depend on how your child advances along two basic interwoven processes—separation and individuation. The separation-individuation process and its derivatives are highly evolved in humans, and are mediated by, but also at the same time, influence the structure of the child's evolving brain. This process is undeniably central to the development of the mind and what it means to be human, and there are as many variations on how this works as there are people.

Individuation is the process of becoming a free-standing person with unique qualities through internalizing and reorganizing what is learned from identification, imitation, learning, and other means. But to become one's own person with a strong claim to one's own self, he must also differentiate and separate from his parents.

Following the intense bonding of early infancy, it is not easy for a child to undertake this difficult process. The capacity to feel secure enough to undertake separation itself and tolerate, if not enjoy, being a distinct individual, in itself requires a degree of self-reliance and awareness that may not have yet formed sufficiently.

Experts have long understood that infants attempt to master the early days of this challenge by inventing a clever halfway measure to assist their transition—an attachment to a transitional object. It is often the child's first toy. It is a universal occurrence. Many children do not invent an obvious transitional object, either because they are prewired with different brain mechanisms, have other typical ways of coping with the challenges of separation-individuation, or, more rarely, because their development has not advanced well.

Here is how the child's brilliant invention works: The child chooses an object—usually a soft and cuddly thing, but it can be anything he can hold and feel. He then endows it with all sorts of qualities and magic—especially to give a sense of safety and calm, like Mother. Now—presto—he has her own private exclusive companion that protects and supports anywhere, anytime, unlike Mother, who sometimes disappears. We also call it transitional because it is neither human nor object, but something in between, and it also helps in an important transition. He may bestow it with a name chosen because it is stable, special, and unique among all other objects in the child's world. Some children go further later on to invent imaginary friends in the same way. *Harvey*, the 1945 play, 1950 film with Jimmy Stewart, and TV show with Art Carney, is an example of this phenomenon.

Here's an example: "Lambie" or "Blankie" is the child's constant personal companion and an important talisman providing safety in the face of the real and imagined perils of her ever-expanding world. She carries this new companion for comfort, support, safety, and soothing, vigilantly keeping track of it and counting on it to reassure and calm her when she is upset. She guards and treasures it, as she takes it everywhere. It enables her courage when facing the mysteries and dangers of the dark night. Clutching Lambie, she peacefully drifts into sweet sleep. It also fortifies daytime explorations into new unknown places.

Later, as her brain centers mature, she realizes that Lambie is not a living thing, but she still nevertheless needs its magical powers and can set aside this knowledge, much as we all know that a movie is pretend yet ignore this knowledge while we become engrossed in it. He may keep this magical invention indefinitely—through college or even later, or at least in her memory, and continue to draw a special comfort from it.

This experience is the child's first encounter with the specialness of play. The child's abilities to partially suspend reality, form a meaningful system of perception, action, thought and feeling, and use a fantasy symbol to make herself feel safer, underlie later play, creativity, and appreciation of others' creations, and the ability to imagine a future and plan for it.

In addition to helping the basic separation-individuation process and its derivatives, it provides opportunities to develop and practice other important human qualities as life goes on. It is the child's major creation, and she fully owns it. She will repeat this many times in many forms over a lifetime, but it is most dramatic when it first appears in early childhood. It later evolves into a complex and rich part of human life.

MORE ABOUT 'WHAT'S REAL'

Relational Artifacts. If present trends continue, parents might soon be highly interactive novel digital media devices in babies' cribs in their zeal to jumpstart intellectual development.

Dr. Sherry Turkle and her MIT colleagues have been breaking new ground in the study of human/machine interactions. These pioneer scientists have seen how elders and children become easily attached to relational artifacts—interactive computer-based dolls programmed to show and vocalize "feelings" and even respond to touch and tone of voice. Young and old alike nurture these humanoid robots as if they are alive. Children struggle to understand the differences between these digital objects and actual living creatures, and sometimes regard the two as interchangeable.

These mechanical pets are helpful to the lonely elderly. Dr. Turkle reports how the elderly in nursing homes enjoy the opportunities for supportive interactions with relational artifacts in spite of their (presumable) awareness that they are mot real.

The humanoids are coming. So would it be OK to place one in the crib? I believe not. When you are a very young child, it is one thing to talk to your own invented transitional object or doll that you animate yourself, powering the moment almost entirely with your own imagination. It is quite another to have the doll talk back in a way preprogrammed by strangers and in a stranger's voice, or even mother's voice. It is one thing to make your doll move. It is quite another to have the doll move on its own. The very act of inventing the transitional object is a major creative and imaginative act. The magical qualities a child deposits in it are directly based on the child's own needs. But a relational artifact comes already invented by someone else with someone else's magic. Moreover, it is preprogrammed with magical qualities of speech, movement, and responsiveness feigning human mutuality. How much room does that leave for the child's own inventiveness and creativity? Even if the child gains intellectually, what kind of person will he be?

As relational artifacts replace the non-interactive transitional object, how would their interactivity affect separation-individuation? Since this process is crucial to the development of socialized, productive, and creative people, do we really know what long-term impact we may be causing on the future of individual development and ultimately on the future of human society? Could the child become a less attached, less sensitive, less spiritual, less responsible, less moral, or less compassionate person? Less creative or developed aesthetically? Less human? We can only guess at the possibilities: addictive or other unwelcome behaviors or deficits, personality disorders, over-reliance on machine intelligence. Will healthy Socialization and Family Relationships suffer?

A relational artifact embedded in a cuddly stuffed animals may appear safe enough. But it may not be. It is programmed to behave like it is alive, but it is not. It is constructed to appear authentic, but it is a facsimile. It may share some qualities with living things, but it lacks life. It may appear to feel or to care, but it does not. It is not alive. It is a fake. Wouldn't it be deceitful to pawn it off on a very young child who may not understand the difference?

When we replace essential human-human contact, or even human-pet interactions, with human-machine contact, are we confusing our very young child about the nature of life itself? Are we interfering with early appreciation of how living things are precious? Would some parents or institutions misuse these mechanical toys to baby-sit or abrogate their role? To control and brainwash? Would your child grow up more likely to take instructions people who act inhumane?

There are more questions than answers about how relational artifacts might influence early development. We obviously know little about this crucial area. But what we don't know, and especially what we don't know we don't know, can hurt us. Answers are beginning to come in, but crucial scientific work is slow and laborious.

Bottom line: wait. We need more research to determine how interactive media can be made safe for children who are undergoing rapid maturation of their brains and development of their minds. What might seem like a generous gift intended to benefit a young child may turn out ultimately to be an unintentional hazard. Experts just don't know how such an experience might ultimately influence cognitive, social, moral, and aesthetic development. Wait until you know more about the benefits and risks and safe use of digital objects and computers with very young children. Be wary of those who endorse these objects.

Thwart those who would have your very young child embrace too fondly intelligent toys, and prudently limit exposure—both in total weekly time and in number of sessions each week—until we know more about their long-term effects. Do not leave the very young child to play alone with them. Be especially alert in this age group to how long and how often your child interacts with these. For the time being, you will probably be better off encouraging appropriate play with gentle pets. Such interactions between living things are complex, spontaneous, non-stereotyped, and provide a back-and-forth that is authentically mutually warm and affectionate. It is what the human brain has been prewired for.

Possible benefits. While the effects of relational artifacts on very young children may be controversial, it is easier to envision how schoolchildren and older youth and other age groups may benefit from educational interactions with them. Later on, once the child is able to tell apart fake and real, it can be an entirely different situation. Nevertheless, we hope that the older child can collaborative thinking and joint exploration, learning, and teaching that only humans can provide.

Parents can start teaching the child the difference between real and fake after the age of about three. In her technology-saturated lifetime, a child is likely to encounter many simulations that will strain the distinction between fake and real, and it may be crucial to start teaching this distinction as early as possible.

Relational artifacts, especially when used in the presence of a helpful adult, can be used as educational tools. However, they must never take the place of relationships with living things. For instance, a five-year-old can prepare for a new baby brother as he and Mother play with a baby doll artifact that requires care and responds to affection. Or an older child and his family may play at caring for and training a responsive relational animal artifact in circumstances where actual animal care is not possible. Or playing with a relational artifact doll that models how to interact with medical personnel and what to expect during hospitalization, preparation, and recovery from a surgical procedure may be reassuring for a hospitalized child and help her and her family adjust to hospital life and routine.

Moreover, like an artificial limb or a seeing-eye dog, media may become acceptable accessories to some intellectually, physically, or emotionally challenged children. For example, an interactive media device that can recognize speech could translate into signs for a hearing-impaired child. Or a child challenged by an autistic spectrum disorder may initially relate better to the more mechanical aspects of a relational artifact, but these could be programmed to gradually increase their "humanness" and random behaviors. We already know that autistic kids gain substantially from portrayals of mechanical objects with human facial expressions. Or play with an intelligent, appropriately trained "therapy doll" companion may provide supportive real-time cognitive behavioral interpretations to a depressed child. None of these would replace human interactions, they just complement them.

The younger the child, the more a parent needs to use relational artifacts within a human, assure that it remains in the realm of play, and mediate the experience to assure intended use. Clarifying the distinctions between animate and inanimate would be the essential task for the grown up. With this caveat, applications for older children might be quite helpful when designed appropriately. Those who study, design, and program relational artifacts for older children would have wonderful opportunities to create effective strategies that would be appropriate and helpful to children's developmental needs.

"We have stepped into a new dimension. Our ability to manufacture fraud has exceeded our ability to detect it." In Japan, where robots are needed to replace an aging workforce, there are thirty-two robots for every 1,000 manufacturing employees. In other arenas, robots spoon-feed the elderly, and those with human-like faces can respond appropriately to words like "love" or "war."

We know little about the potential effects of intelligent toys on children who cannot yet clearly separate fantasy from reality. But we can take some comfort that people have been imagining the challenges of interacting with simulated beings for centuries. Their ideas may be especially informative here, as we try to predict the roles that such objects might play in our children's futures.

We have already invented a rich lore about them that will give us some clues. Relational artifacts go back quite a ways. The Golem is a fictional humanoid relational artifact in early Jewish folklore. According to tradition, it was made in AD 270 in Babylon by Rava, a Talmudic scholar said to have used the mysteries of the mystical

Kabala. The Golem served various roles, but mostly as a servant and protector. *Pinocchio* (ca. 1882) by Carlo Collodi is the beloved novel for children which features an animated, boyish marionette.

With the industrial revolution, fictionalized narratives about how human beings interact with intelligent machines have captured the popular imagination. All sorts of talking, blinking, and even wetting and walking mechanized dolls have proliferated over the past century. In the 1990s, kids interacted with small hand-held digital gadgets that simulated a growing baby and communicated its needs via an LCD screen.

Filmmakers especially, have explored this area's complexity, probably not coincidentally, since the very essence of film is the mechanical creation of a magical illusion that challenges us to ask, what is real and what is fake? Woody Allen's *The Purple Rose of Cairo*, where audience members and movie characters actually cross into each others' worlds, and the cultish *Rocky Horror Picture Show,* shown ritually late at night, where audiences play dress-up and interact as the film's characters, both explore the grey area among film, illusion, theater, and real life.

The simulated world and its creatures are not always friendly. There may be grave consequences to fooling with Mother Nature. Science fiction films like *The Island of Lost Souls (*1933) based on H. G. Wells' 1896 *The Island of Dr. Moreau*, the Frankenstein stories, the Chucky movie series, the Borg in *Star Trek,* and HAL in Stanley Kubrick's *2001: A Space Odyssey,* all portray the perils of relational artifacts and imply that their makers should be more careful. In Lisberger's *Tron* (1982), a human is actually swept into the circuitry of a supercomputer where he battles digital forces that would destroy the world. Of course, there are films where humans are transformed into degraded, robot-like simulations of their former selves, like *The Stepford Wives* (1975, 2004).

On the other hand, not all relational artifacts are bad. Some may actually be helpful and form meaningful relations with ordinary humans, as do C3PO and R2-D2 in Lucas' *Star Wars (1977).* Other simulations, like *The Wizard of Oz (*1939, based on L. Frank Baum's 1900 book), more closely resemble the way children actually imagine and use dolls in normal play.

Or the picture can be more complex. Both *Blade Runne*r and Kubrick's and Spielberg's *A.I.* intelligently explore the complex benefits and hazards of relational objects configured as closely as possible to virtual humans. Hauntingly beautiful and sad Rachel, and the noble, poetic android leader, Roy, are portrayed lovingly in Ridley Scott's classic *Blade Runner.*

Andrew Niccole's lighter *Simone* (2003) deals extremely well with these serious questions. Viktor, a failing film director, replaces a temperamental human starlet with a computer graphic avatar of a beautiful woman capable of the non-verbal nuances of being human, such as subtle postures and facial expressions. Viktor keeps the deception going even as Simone becomes a pop-culture superstar. She explains not making personal appearances, "I relate better to people when they're not actually there." Like other films, *Simone* argues convincingly that intelligent non-living simulations seriously challenge us to understand how they change us. In *Simone*, one character states, "I'm starting to feel like I don't exist," but another entered rehab to kick her drug habit because "Simone inspired me." So who is influencing whom? Viktor pleads, "I made her," but others see it differently—"She made you" successful and rich.

Viktor states, "If the performance is genuine, it doesn't matter if the actor is real or not," and, "We have stepped into a new dimension. Our ability to manufacture fraud has exceeded our ability to detect it." Such statements can easily describe the media merchandising already capturing very young consumers.

Near the very end of the movie, his teenage daughter admonishes Viktor not to play fast and loose with the boundary between fake and real. She says sweetly, "We're fine with fake, as long as you don't lie about it." In other words, as if to say, "Dad, teach us kids the difference between fake and real when we are old enough. Until we know the difference, we trust you not to deceive us, or allow others to deceive others."

But true to its sarcastic core, *Simone* ends with a reassuring wink, as if to add scathingly, "but don't worry, Dad, I don't really care that much. It's really OK for us in the mass media to lie about our deceptions. No one notices or cares how much we are already lying anyway. In fact, people like what we do and vote 'yes!' with their wallets at the stores all the time, and always want more and more." The teenager, a quick study, she might have added, "And, oh yeah, Dad, don't forget all the kids out there. Did you know that we can actually get them to know what they want as early as—get this—two or three, and then they'll pester their parents to buy? We might as well start them out real early, Dad."

The jury will be out on how to best use interactive media devices endowed with human qualities such as relational artifacts for quite a while.

Technological advances are already forcing us to rethink the lines between life and death, living and inanimate, toy and non-toy, and human and non-human. In a broad sense, the debate about interactive devices might be placed together with questions of cloning, stem cell research, biomedical devices, medical definition of life, and artificial intelligence, going to the heart of ethics and the type of values upon which we have based our society. In our children's technology-rich world, clear landmarks that we have taken for granted to orient us and map our everyday realities are becoming blurry.

Our children will face challenges to the social, ethical, and moral principles that define them as human and enable them to live together. For instance, if God made humans in his own image and men made robots in their image, are robots God's creatures too? Do we refer to a relational artifact as an "it," or as a "he" or "she"? Here I will awkwardly use what seems to feel intuitively right, but I am aware that this problem is at the very crux of the matter. I believe that we would use "it" when the object is not close to our experience, much as a child calls her new doll an "it" in the toy store, but changes to a human name and refers to it as a "he" or "she" when it comes home.

Chapter 12: The Future I: Interactive Media Play Therapy and Other Clinical Applications

This chapter is intended for professionals working with children. Others may also benefit from the frameworks given here and learn about play therapy, its importance, and new applications.

Play therapy is a traditional and effective way to understand and communicate with children. In this chapter, I will lay the basis for how therapy works in general, and how it works with children in particular. You will find a demonstration of the role of play in learning, developing, coping, and healing. Then you will see the practical opportunities that interactive media offer the practicing therapist. I will then describe my own experiences using interactive media in my practice and give many examples.

WHAT IS PLAY THERAPY?

To understand play therapy, we must first understand therapy in general. For youngsters, as for many adults, the main task of treatment is to get rid, as much as possible, of obstacles to their development and functioning. Psychotherapy, counseling, interpersonal, or talk therapy, is often an important aspect of the treatment program. When done correctly, such help can be invaluable.

Here is my way of understanding therapy that will serve our purposes here. Let us review its basics. Psychotherapy, in its traditional dynamic or interpersonal form, is based on psychoanalytic theory and practice. It is this theory that still provides our most careful and comprehensive understanding of how the human mind develops and works. Although a confusing plethora of so-called therapies and economically-driven knock-offs proliferates, it is traditional psychotherapy that is the richest, most humane, caring, intelligent, and often most effective form of psychotherapy. I further believe that many of its principles are being validated as we get to know the brain better.

I like to think of psychotherapy as happening in the minds (or between the brains) of the patient and doctor in a special space that they create together apart from ordinary life, much like the space children create in their play. At its best, it is like a safe, supportive, emotional, and intellectual enclosure, structured within a professional collaboration that encourages development and healing to flourish. It is a special space that every therapist and patient creates together to suit the needs of the patient. Both bring an initial expectation and hope. The therapist encourages further hope as he psychologically "holds" the patient, as a mother might an infant, through professionalism, expertise, and the supportive posture of a healer. The therapist also provides emotional safety, boundaries, basic scaffolding and other formal features of the "holding environment." The therapist intends that this "holding environment" will enable a patient to feel safe, supported, and hopeful. I believe that almost no healing is possible without this space.

The patient responds as a co-member of the "therapeutic alliance." Within this special space, the patient and therapist collaborate and problem solve together as the patient tells about himself and those around him. For some, trusting enough to get to this step, the risky act of hearing oneself reveal to a non-judgmental listener, is itself monumentally therapeutic. The patient may tell of just one event, a relationship, or his whole lifetime. The therapist and patient then reassemble the facts and retell the story together so that the patient understands more clearly the damaging beliefs he has been carrying. Together, patient and doctor collaboratively rethink and reframe the patient's misperceptions, misbelieves, and unhelpful behaviors. The patient recreates himself in a more forgiving, positive, and loving light. Sometimes, the doctor uses medication to adjust biological events and patterns in the patient's brain that lead to major problems in thoughts and feelings.

This healing partnership does not exploit the patient; and is neither a friendship, nor is it egalitarian. Helping the patient to become a necessary partner in this process, the therapist knows from training and experience what usually works and what does not. Often, the therapeutic space extends into the patient's daily life as he carries it in his mind and remembers its supportive and helpful aspects. Eventually, it can become woven into the very fabric of the patient's self and, long after the sessions have ended, remain an inner resource for nurturance, calm, self-esteem, gratitude, forgiveness, compassion, creativity, and strength. Psychotherapy can clearly re-pattern the way the patient's mind works, and these changes become embedded in the very circuitry of the brain itself. Sometimes, like the "butterfly effect," a small change can lead to large overall improvements.

These principles also hold in the psychotherapy of children, where the therapeutic space is more clearly the same special space of children's play. Play in the doctor's office is different from play elsewhere because it is play inside the therapeutic space. Through their play, children act out their stressors, strengths, fears, talents, loves, ambitions, angers, and otherwise reveal themselves within the collaboration with the therapist. Play therapy is a means to get to know the child, communicate and build a therapeutic partnership with her, and work to achieve specific goals. Most therapists have a variety of materials they have offered children for years. How a child selects his play or art materials and activities is itself diagnostic and therapeutic. Children often find their own preferred toys or art materials from among these, or sometimes bring their own.

In any case, therapists can select play materials and activities deliberately and carefully to enable conversations with children and boost specific therapeutic aims. Therapeutic play interactions are complex and usually achieve several goals at the same time. The child can work alone, or the therapist may watch while they talk, or the two may play together actively, creating a game as they go along. Play may be the sole activity in a therapy, may be temporary, or may be part of each therapeutic encounter. The skillful use of play in therapy requires sound grounding in child development, open-mindedness, good judgment about how active to be and what to do (or more importantly, not do), and the ability to recognize and appreciate the child's imagination, creativity, and experience.

TRADITIONAL PLAY THERAPY EXAMPLES

Two-year-old Dana survived an automobile accident but has become withdrawn. In my office, she spontaneously selects two toy cars and crashes them into each other repeatedly, thus starting a collaborative therapeutic play conversation leading to her improvement.

Eight-year-old Matt, demoralized by a reading impairment, learns to play chess with me and slowly gains confidence in his intellectual abilities. He comes to believe that his reading impairment does not make him a "retard."

Comment: These children have worked out ways in traditional play therapy to remove developmental obstacles.

A six-year-old boy plays doctor with dolls under the direction of a therapist in the hospital playroom. He directs the therapist to play nurse. He then becomes the surgeon. He thereby asks for help in learning and mastering his anxiety about his illness and pending surgery and recovery. He continues a version of the play as he recovers in order to understand his experience and put it in perspective.

The four-year-old sister of a newborn spontaneously pretends to be a nurse. She later selects a baby doll and plays "mom" to her new baby, talking to her baby and otherwise imitating her mother's behaviors as she adjusts to an important change in her life and explores her own future motherhood. She directs me to play doctor or father or store clerk as we collaborate in working out a variety of her concerns.

A traumatized six-year-old witness of a school shooting draws her version of the event over and over and cannot move past this subject. She describes her feelings at length. I offer her alternative interpretations of her role in the event, and she improves.

An agitated boy and I walk, because he cannot tolerate sitting face-to-face for a long time. We also play catch on a playground.

A clumsy and withdrawn seven-year-old gains confidence and an improved body image by shooting baskets non-competitively with me. We each support the other and keep a running total of how many baskets we make together.

Elaborate play as the main vehicle of therapy with a child suffering from bipolar disorder and a severe learning disability.

I determined that Charles is an impulsive, aggressive, and isolated five-year-old, suffering from bipolar disorder and severe learning disabilities. During the course of years of treatment (that also includes family therapy, school consultations, and medication), Charles uses play therapy to cope with his challenges and advance his development. He invents and reinvents a succession of clever games. For example, around age nine, he creates and revises an elaborate war game that he continues to play for about a year.

Typically, he builds a bunker around my chair from piles of sofa cushions and pillows. He alternately assigns me the roles of medic, superior officer, buddy, lookout, or underling as we face an imagined enemy together. We plan and carry out defensive, reconnaissance, and offensive operations together, some quite complex. He directs the use of a variety of weapons, communication and transportation, and computer equipment. At other times, we agree that one of us stands guard while the other sleeps (and snores and has nightmares that he tells about), one saves the other's life, and we each tell our bunker buddy about our imagined family back home, hopes for the future, and other imagined personal matters.

Charles has found in his therapist a helpful companion in his lonely and conflict-laden life, who appreciates his creativity and inventiveness. He slowly learns to appreciate his own considerable intelligence and strong moral sense and makes friends. He finds ways to control his aggression and impulsivity as the two "fighters" carefully analyze each battle situation, plan their strategies, and exercise restraint and best judgment in a gentle therapeutic space marked by a warm attitude of patience, acceptance, mutual respect, and cooperation. Later, Charles enters high school and plays varsity football, acts in school plays, and develops a special interest in the military history of the American Revolution. Upon graduation, he enters college with the hope of majoring in history and becoming a special education teacher or a psychologist.

Comment: These examples provide glimpses into the nature of play therapy and the power of play as an organizer and facilitator of rich healing and developmental experiences for children. Although experienced therapists develop their own play styles and materials, the same basic principles and skills must apply to all forms of play therapy. There is no reason why electronic interactive media cannot provide opportunities for play therapy.

INTERACTIVE MEDIA PLAY THERAPY (based on a paper presented at the Annual Meeting of the American Psychiatric Association, San Francisco, CA, May, 2009.)

Kids have been bringing CD players and MP3 players to my office for years. Often, the parent asks the child to leave it outside. An MP3 player or equivalent, flash drive, PDA, or cell phone-based storage, in their current configurations are digital music storage and playback devices, currently used mostly for entertainment. The MP3 player enables a child to create a portable personal music library, revealing the child's esthetic development and organizational style. Its most important feature, however, is that it enables selection and easy playback of music.

A few years ago it suddenly occurred to me—what an opportunity! What better way to access feelings than through music? Disclosing, sharing and conveying feelings via the music selected by a patient can serve therapeutic aims. I understand music, and it is an essential part of my life, so this became a no-brainer from me from then on. Children already use interactive media devices as toys and often spontaneously ask me if they can play computer games or show me things online on my office computer.

INTERACTIVE MEDIA PLAY THERAPY

- First, needs to be accepted by family as a legitimate professional activity
- Enables specific therapeutic interactions within an overall treatment plan.
- Enables breaking the ice and establishment of a collaborative therapeutic alliance.
- Enables direct observations of how the child functions socially.
- Enables patient self-disclosure.
- Enables work on excessive media use.
- Enables rich, ongoing therapeutic interactions that enable support and cooperation.
- Is a legitimate play activity within a therapeutic space.
- Equipment is used by children as play materials as well as the visual and audio information it displays.
- Offers therapeutic opportunities not as accessible by other means.

When a child brings in an MP3 player, I offer to listen together via the music system. The child connects his MP3 player to my system through its earphone jack. Immediately, the anxious child is offered relief. It is a way to encourage the patient to create the therapeutic space with me, a place for being with a kind doctor, who is not judgmental, who accepts and understands kids, and who appreciates and respects their world; a doctor who will meet the child at least halfway.

Kids often bring an MP3 player and play a selection. I listen carefully. When I do not follow the lyrics, we find them online and view them together (see below). I often reveal the feelings the music evokes in me and appreciate these as ways of showing how the patient might feel too, since music is a great way to express feelings.

Similarly, with the intention of reaching children, I began to make the online computer available for use as a toy in my office. I decided that I could offer kids use of interactive media devices as long as I used the traditional conceptual framework of play therapy responsibly and thoughtfully.

Interactive media devices described here include the online computer, MP3 player, video game, CD player, flash drive, Blu-Ray DVD, and mobile or smartphone. These toys offer a range of play activities from solitary to highly interactive. Perhaps the term "Interactive Media Play Therapy" would differentiate these interactive media devices from traditional play therapy. The term is applicable to the Internet and other current media now in use. Moreover, it could stay useful as technology brings other applications for digital interactive play therapy.

As play materials, interactive media devices can provide unique clinical opportunities for observation and for activities that enhance communication; ease tensions, and a variety of other therapeutic interactions. In many cases, portable devices serve as ways to connect to others. But they also sometimes serve as security objects that relieve the anxiety of aloneness in an over-stimulating, difficult, and chaotic world.

The multi-function mobile phone and online computer can provide many opportunities to reach well beyond the office walls to join the child in a variety of real-time and real-life social, esthetic, and intellectual experiences. To children, the mobile phone is more of a toy accessory that defines a child's individuality and status. Its communication value increases with time; text messaging falls in the category of instant messaging described elsewhere.

Setting up an online computer for interactive play therapy is straightforward with easily available hardware. An MP3 player or CD player can be connected to a sound system or the computer with an easily available "Y" audio cable. DVDs and CDs can be played on most computers.

In my small office, the child holds the keyboard on his lap, sitting on the couch, while I keep my laptop on my lap, sitting in my chair at a right angle to the child. I pull a 22" LCD display from behind a couch, place it on a small coffee table, and connect it via cable to the laptop. This completes the triangle. This arrangement also makes listening to CDs and viewing DVDs possible, giving the child opportunities to show and tell about recitals, sports activities, graduations, or school plays if appropriate. In spite of its sophistication and cost, the equipment is in fact only play material and should be offered with customary thoughtfulness, especially to a child who has sensorimotor challenges or who can become too easily overstimulated or destructive. Obviously, if the child is too active, aggressive, or might abuse the equipment, he cannot use this form of play.

Whatever the device, its use must promote therapeutic interactions. Human-interfaced computer systems are being developed that might be useful for interactive media play therapy. For example, Microsoft Corporation has recently introduced a system configured like a coffee table with the screen as the tabletop. Several users are able to look at each other as well as the screen, while the computer identifies each individually. These will allow several users to interact with each other and with the machine and with the online world at the same time.

As with other modes of play, selecting and using interactive play must always be a deliberate, clinical decision that is part of a plan to achieve therapeutic goals. I generally regard use of interactive media devices as I do other toys in play therapy. Since I generally store play material in a cabinet, I also keep the large monitor out of sight. For some kids, interactive media play is best at the beginning of a session, while for others closer to the end. As usual, play duration and limits are all deliberate clinical decisions. Parents often need an explanation up front about play therapy with children and interactive media play.

Overall, the therapeutic task must remain assisting the child to do away, as much as possible, with impediments to her development. The therapist must monitor its helpfulness and reevaluate its use on an ongoing basis. Interactive media play therapy is not a substitute for traditional therapy, but it does seem to have a place in therapeutic space. Except for rare circumstances, all digital electronically-mediated interactions must include a large interpersonal component between child and doctor.

Once I decide to attempt interactive media play, I might say to the child, "One way we can get to know each other might be to go online together right here in the office, if you want to." Some children are too inhibited to agree, as they would be to any form of play. Others agree, and I set up the arrangement described above. Some are surprised and delighted, since no grown-up has ever actually offered to go online with them. As with all play therapy, this is all grist for the mill.

The Future I: Interactive Media Play Therapy and Other Clinical Applications

Figure 12.1: Therapist and youngster interacting with each other in a triangular setup of an online laptop, wireless keyboard, and a large display.

Remind the child that you would like speak together and look at each other as much as possible, because this activity is part of working together to understand and make her life better. Use the browser's zoom function (CNTRL +) to enlarge the display image.

To set up a triangular arrangement at home or office, here are some suggestions for Windows XP users:
1. Keyboard and mouse: Plug another keyboard or a wireless keyboard with a built-in mouse to a USB port.
2. Monitor: 22" or larger is best. For easier distance-reading you can also adjust Windows' font size to extra large (control panel> display> appearance). Audio can come from the computer.
   a. Better option: Keep your present monitor and simply turn it and rearrange the seating.
   b. Get a new monitor. If your desktop has the proper output, or if you are using a laptop, most times you would succeed by just plugging the new monitor into the proper port. When in doubt, best consult a help line. The image on your added monitor will probably have the same resolution and shape as your primary monitor, so it may not show its best. A video graphics driver or accelerator, or switching cables and changing the

display settings (control panel>display>settings) may improve the picture. But remember—this is play, and toys do not have to be perfect.

I then say something like, "OK. Now we can play. You know, Jake, this is part of what we will do here together, so looking at each other and talking together is very important to me." Most children agree, but often perfunctorily, and need to be reminded often.

After teaching about using the wireless keyboard, I usually give the child free reign, as with other forms of play. Since most children's only experience with the computer has been solitary, they may not be used to simultaneous interactions with an adult. The therapist must be sensitive to the child's need for gentle patience as he learns that the interaction is indeed meant to be therapeutic and social. When appropriate, I attempt to keep a conversation and questions going. Occasionally, I remind the child to pause, and we talk about what we saw and the relevant whats, whens, wheres, and hows. Sometimes, I establish firm limits, such as the proper treatment of the equipment, or I might give gentle five-, and then two- and one-minute warnings to children who have difficulty stopping.

Interactive media play helps break the ice. In general, children can more easily warm up and make use of the doctor in his office as their special therapeutic space, or "holding environment," when offered familiar play material. And what is more familiar today than the online computer or MP3 player? When a doctor takes their music and online activities seriously and reacts to it with genuine interest, children feel understood and respected. It is easier to trust the doctor. This play is often helpful in the early phases of therapy in establishing the therapeutic space and a sense of collaboration.

Direct observation of children in their natural environments provides a wealth of helpful information. In some situations it is essential to obtain a firsthand view as part of a clinical evaluation. For instance, as part of a complete evaluation, a child psychiatrist sometimes visits the school, talks with teachers, social workers and psychologists, and watches the child in the classroom and on the playground. The use of Webcams connecting the school to the psychiatrist's office via the webcams could make such conferencing and direct observation more convenient, affordable, and practical. Conceivably, such use of available interactive media may make possible voluntary displays of interactions of family life in the child's home.

Certainly, the use of an online computer in the doctor's office provides a wealth of opportunities to observe the child's personality and to learn about and intervene in her social life and other developmental areas. How the child approaches online relationships, how readily she trusts the therapist, what interests her, observations of her impulse control and attention span, her facility with keyboard and mouse, and how she interacts socially online are some observational data. Special HIDs (Human Interface Devices) can similarly assist the psychiatrist's work with hearing or otherwise challenged children. Other digital electronic devices like MP3 players can provide opportunities to learn about the child's inner emotional life.

Much like other play materials and toys, the online computer provides its own special opportunities for specific therapeutic goals, illustrated by examples below. The triangular configuration in the office allows rich, ongoing therapeutic interactions, as both doctor and child see each other and the computer display, and both control the cursor and input text. This arrangement offers the child opportunities to practice reciprocity, collaborative thinking, mutuality, respect, courtesy, kindness, patience, and to delay impulses.

Children appreciate the opportunity to share directly a part of their lives. Visits to game sites can break the ice and provide opportunities for an initial engagement. Some preadolescents and adolescents share social media pages and instant message contacts with the doctor in a manner they would not with their parents. This openness allows therapeutic conversations about their social concerns, sexuality, privacy, and other matters. Other children benefit from the shared excitement of learning and discovering with the doctor, for instance on Google Earth or YouTube. Other, more specific goals may include diagnostic observations of children who spend excess time online or who have major conflicts with their parents about the computer and online time.

A therapist may have his own Web site or other interactive means that a child may visit and then sign on to his own unique pages, even using a mobile device. This arrangement may allow at least "touch base" that revitalizes the therapeutic space. Similarly, a child may carry in her interactive mobile device images of the therapist. Other opportunities can include seeing a video of the therapist repeating therapeutic or other information, medication warnings and questions, suggested links, and continuing treatment via a text thread or a real time chat or Webcam teleconference. Such applications would be reminders that extend the power of therapeutic space beyond the walls of the therapist's office.

Because so little is known about their impact, I cannot recommend routine clinical use of interactive media before more systematic studies refine our understanding of their benefits and pitfalls. I recommend that only prudent, properly trained and credentialed professionals, already well versed in the theory and practice of child development, play therapy with children, and therapy with families, extrapolate from the contents of this book to their own clinical work.

INTERACTIVE MEDIA PLAY THERAPY EXAMPLES

Six-year-old Andrew complains that he has no friends and sits alone during recess. His parents report that Andrew has difficulty taking turns and respecting his classmates. He also complains of boredom.

Andrew asks if he can play on my computer. I decide that interactive media play might be helpful. Andrew takes me to an online Mario-like game designed for one player. At first, Andrew plays alone, and I cheer his successes and console his failures. But Andrew soon complains that he is bored. He agrees that this is exactly the same he feels on the playground during recess.

Andrew invites me into the game. This sequence happens in several sessions. We work at learning to collaborate with each other. Andrew's tendency to "hog" diminishes slowly, and he eventually asks to play together, with each controlling the same cursor—a real challenge.

I tell Andrew in many ways how his behavior on the playground and elsewhere is similar, and Andrew tries to be more cooperative. As his reputation slowly changes, he becomes more and more welcome into his peers' play. He has discovered the pleasure of collaboration.

Comment: This is an example of how the online computer enables Socialization. Children have made interactive media devices favored toys in their world. When appropriately offered, interactive media play therapy is a helpful extension of a time-tested method and enables breaking the ice, observation, self-disclosure, and therapeutic interactions in ways not as accessible otherwise.

Children online are children becoming what they can be. In order to understand the hows and whys of children's usage of the Internet and other interactive media devices, we must first understand as fully as we can the importance of children's play.

Eight-year-old David makes a fashion critique about my computer desktop when he recommends that I dress it up and improve its minimalist appearance. He thinks I should consider making it more personal. I acknowledge his point and ask for suggestions, which we try out. We then discuss how his preferences may differ from mine and agree to disagree.

Comment: Older children undergo rapid and powerful development in both their individual and social group identities. They work on staying unique and special within the family and social group, while at the same time craving acceptance. Their social groups can be extremely demanding and critical, even tyrannical. Customizing the appearance of their cell phones, computer screens, and social network sites are opportunities for development in this area. Their uses of interactive technologies reflect these normal efforts to find the correct balance among these seemingly contradictory needs.

In this case, David is demonstrating that he is developing in this area and is opening a therapeutic conversation with me.

The MP3 player is a lifesaver. Seventeen-year-old Sara, a first-semester senior, is not prepared to face the imminent prospect of leaving home for college. Her schoolwork suffers because of fatigue and poor motivation and concentration. She is losing weight. Her social life suffers as she withdraws, and she has no fun except when she drinks or smokes weed. She starts experimenting with self-mutilation and begins to associate with a sexually promiscuous crowd that abuses substances regularly. Her parents, in denial and consumed with their own chaotic lives, finally agree to send her to a psychiatrist on the advice of the school psychologist.

It becomes quickly clear to me that Sara suffers from worsening depression and is fast approaching danger from suicide. Sara initially resents being in my office, keeps her coat on, and sits as far as she can from me. She mostly sneers and avoids eye contact and speaking. When she does speak, her thinking seems jumbled and slow.

In an effort to initiate interactive media play, I show interest in the MP3 player Sara is carrying. Sara brings it to next session and agrees to share her music with me, so I hook it up to my sound system. Sara plays music that expresses loneliness, loss, longing, and sadness. Some music is wild and violent with pain. She quickly appreciates my interest in both types of music and thoughtful responses to the lyrics. She agrees to repeat lyrics I cannot hear or discern. She becomes less withdrawn over several sessions of listening together. Sara becomes more hopeful. She widens the conversation to include her feelings about her home life.

Sara begins to trust me sufficiently to start medication. As her depression remits quickly, she says she finds coming to see me helpful and wants to continue. However, her parents "do not believe in psychiatry" and are upset when they discover that some sessions are spent just listening to music, proving to them that all psychiatrists are charlatans. They refuse to allow Sara to continue office visits or the medication, although they have the financial means. They do not respond to my attempts to reach them and to warnings that Sara is at high risk for severe depression, substance abuse, or suicide without ongoing psychiatric help.

When Sara stops coming or responding to my phone calls, I send her an e-mail urging her to go to a local free clinic. I alert the school psychologist to keep a close eye on Sara and to also urge her to go to the clinic. Happily, Sara does go to the free clinic to continue medication and psychotherapy. Her depression eventually remits completely, and she is able to move forward in her development and go to college.

Comment: In this case, speed was crucial in allowing the patient to become involved in her treatment before her parents interfered. Thanks to interactive media play with an MP3 player, she was healthy enough to recognize the benefit of professional help and to resume it on her own. This example also shows that the use of interactive media devices is not a magical cure-all. It is but one aspect of evaluating and treating complicated youngsters and families.

An MP3 player and the music she brought into the office made it possible to quickly treat a very ill, resistant teenager. Her choices and reactions to music revealed her state of mind, longings, moods, and identifications. The therapeutic conversation expanded to include the child's evolving esthetic development and rich emotional states that music evokes best.

Eight-year-old Jon is too busy multitasking to have a decent text conversation with David. After logging on to a text messaging service and hailing David for a conversation, eight-year-old Jon quickly logs in to myspace.com to check the latest social news, then jumps to epinions.com to search for a new camera, and then to amazon.com to price it. David's waiting message signal blinks unanswered on the bottom of our screen. When I point it out to Jon, he returns to the text messaging service and reads David's message which says that he thinks Jon is boring.

I point out to Jon how his rude behavior might have hurt David, and how his general tendency to be inattentive in social discourse can demean his social partners. Jon quickly grasps his mistake and knows what he must do. He returns to the messaging service, reengages David, reassures him that he is not boring, and even asks him for advice about a camera. Jon feels good about repairing his relationship with David.

Comment: Jon was suspected of suffering from Asperger's syndrome by his parents. To confuse the picture, he suffered from depression from ages three through six, which isolated him and severely limited his social experiences and opportunities to learn essential nuances of social discourse. Although his depression is in remission, he is still somewhat isolated by his deficits, especially timing—the crucial reciprocal rhythm necessary for affirmation of contact and maintenance of a satisfying sense of mutuality. He leaves long gaps in interactions. His timing is awful. He has yet to be fully attentive within a social encounter to satisfy his healthy need for contact and give his social partner a sense of being valued.

Paradoxically, his restless multitasking online is yet another expression of his desperate, seemingly safer, attempt to escape his loneliness and emptiness. Play therapy within our strong therapeutic alliance enables us real-time shared observation and an immediate opportunity for collaborative thinking about his social behavior and opportunities for remediation.

Seventh-graders Jeremy and Ari made up when Tyrone mediated their conflict using instant messaging. Ari, stressed by events at home, vilified Jeremy to Tyrone in an instant message.. Jeremy, feeling humiliated, lashed back at Ari through yet another message to Ari via Tyrone. Ari quickly regretted his actions, but did not want to lose face by apologizing. Tyrone became upset when he witnessed the online conflict, and mediated between his two good friends. Tyrone helped Ari and Jeremy make up online after explaining their points of view and apologizing.

Comment: The Internet is a place for complex social transactions for older kids. Accessibility, anonymity, speed, and easy dissemination shape the nature of interactions. Things can be said in text and impulsively disseminated more easily than face-to-face or by telephone. The sender can feel detached, but his impact can be just as devastating. Injuries can result easily, but relationships can also be repaired quickly. In this case, Jeremy told me that, on the whole, the Internet was helpful because the indirectness of instant messaging allowed the parties to make up in a manner not possible face-to-face.

Eight-year-old Jane has a disorder that interferes with her ability to make fine social judgments and pick up on subtle social cues. She is generally behind her peers and has been slowly excluded from her old social circle as her friends move along their developmental trajectories, leaving her behind. She has been missing out on the fun of collaboration and improving her skills in this area. As Jane feels increasingly self-conscious, she withdraws more, depriving herself even more of challenges and opportunities to advance socially. She is definitely heading the wrong way.

As part of a long conversation spanning many sessions about what she can do to end up feeling better about herself, Jane proudly takes me on a virtual tour on YouTube into her garage where she was playing with some of her few friends. We look at the clip carefully and noted together how she has become more sensitive to others' social cues and deferential to their needs. I note that she is more successful socially than she gives herself credit for. She comes up with more ideas to improve her social standing. We collaborate in thinking about her strengths as she increases her positive activities with her peers.

Comment: Children easily share excerpts from their private lives with the doctor by using the power of the online computer to transcend time and space, creating opportunities for overcoming challenges and maximizing their development.

Jeff's mood changes were erratic and hard to understand. He is sixteen, enormously bright, but painfully isolated and frightened. Like many adolescents, he is difficult to diagnose because of an evolving and shifting clinical picture. During one session, Jeff mentioned in passing a moment the day before when his mood suddenly plunged, but he could not describe exactly what it was like for him. He then remembered that he had saved a prose tidbit he had written during the episode. He kept such notes for later inclusion in stories he might write. We logged on, and he showed me the note.

Jeff said that it was a poor example of his writing. When I noted how the text revealed a shift from an initial optimistic and well-focused state of mind to a darker state where his writing appeared muddled and confused, he lit up, saying that maybe the text was OK after all. It now seemed better than he had initially thought because he did succeed in conveying the sudden shift in his state of mind. We wondered if this was an unwanted side effect of his medication or actually new diagnostic evidence about the nature of his illness, or maybe both.

Comment: His online folder allowed this child to easily save a private moment and to readily show it to me as we collaborated in thinking together about his difficulties. The text was not just a report of his mood shift, it was the experience of the mood shift itself, literally as close to it as words would get. The online computer provided this opportunity for a diagnostic window and for strengthening the therapeutic alliance.

Brad wonders how he can keep being the person who he is becoming. The eleven-year-old showed me his dilemma online better than he could describe it in his own words. He took me to YouTube and showed a short and charming animated film.

A creative individual escapes the drudgery, dehumanization, and anonymity of his life as a factory worker when he is inspired by his inner gifts to invent special goggles that enable the wearer to see beauty, color, and light in an otherwise endlessly gray, impersonal, and dehumanizing world. The character was proud and excited about making the world a better place. But in an ironic twist, his bliss is cut short as he soon becomes part of the dehumanizing system again; only now he is the loathsome head of the drab, inhumane, and impersonal factory that manufactures his goggles for the mass market.

I grasped quickly that Brad is telling me that he has joined the ranks of thoughtful and talented adolescents struggling with a problem they are now seeing with a newly developed awareness: can he keep his newly found, and yet fragile sense of himself as a unique and alive person who can do good in this world, and at the same time also fit in and succeed in an impersonal, intellectually and esthetically drab and inhumane mass culture that promotes conformity?

Comment: This is an example of how the online computer enables aesthetically and intellectually developed adolescents to access and utilize art to clarify and express concerns that they could not otherwise articulate as poignantly.

His parents think he is addicted, but Henry just loves to chat. They complain that their thirteen-year-old is always online and accuse him of being addicted, and that all he is doing is playing games. Power struggles and conflict create anger and resentment all around. In my office, interactive media play revealed immediately that Henry was actually appropriately social and was mostly instant messaging his friends.

Henry said he does not like fighting with his parents and that the conflict spoils his enjoyment of his online time. He agreed to limit his online time to five minutes online and stop within one minute of my request. Initially, Henry could not follow this plan as agreed, and I had to be firm. Soon, however, Henry became compliant. Henry generalized this change to his home media use after his parents agreed to give him five minute warnings.

Comment: The online computer allowed easy observation and remediation of the child's behavior, leading to helpful understanding and to changes in his and his parents' behavior.

Jake is stuck. At twenty-one, he has been out of high school for three years. Since about age twelve, he had been distant from his parents. He did not work hard in high school, and had only a few friends. After failing his first semester in college, he returned home and held a few entry-level retail jobs, but felt these were below him and quit. For one reason or another, he could find no other suitable work since. He considered a local junior college, but has been unable to motivate himself to enroll.

At this point, he hardly leaves his room and barely speaks to his parents. He rarely goes to a movie with a friend. He is socially and emotionally isolated and is becoming more and more afraid to leave the house. He eats alone and sleeps late into the day. He stays up late doing nothing, according to his parents. Jake's parents claim they want to help him, but also claim helplessness, and have not undertaken serious efforts, while his grandparents indulge and baby him.

As we go online together, Jake amazes me. He shows boundless curiosity and a terrific hunger for learning and an ability to acquire, retain, organize, conceptualize, and recall information. He focuses on a topic, researching it intensely to a point that he would be considered at least well-versed. The topics range from South American soccer and Cajun music to climbing Mount Everest.

At times he listens to Talk Radio from all over the country or watches videos on all sorts of sites. When activity in one area lags, he turns to another running open window. He is excited and feels enriched by all the valuable free opportunities from which he can learn. He has become an expert in certain areas, like LCD technology, and has followed the industry carefully, noting corporate changes and other news. He learns enough about climbing Mount Everest that he could visualize being there, knowing the details of clothing, equipment, and climbing strategies.

Eventually, Jake enrolls in a local junior college and develops a social life.

Comment: Jake's development as a person was slowing down through high school and had come to a stop, except for his intellectual online interests. He was unable to move on into young adulthood because he lacked the necessary emotional and social underpinnings. His sad story is not uncommon, and is the result of a long series of smaller, partial developmental failures since early childhood, not noticed by his parents.

Jake was avoiding life's challenges, many of which he feared he was ill-prepared to meet. He was settling for the exciting and stimulating, but poorly directed and lonely virtual world. In addition, he probably also suffered from an undiagnosed depression for many years.

The online home computer is a mixed blessing to Jake. On the one hand, it offers him a refuge and some Growth Opportunities that may have saved him from a total breakdown. On the other hand, Jake's use of the computer was undisciplined and unguided, as was the rest of his life, and insufficient to move his development forward. This case illustrates how important it is for parents and doctors to understand the child who uses media excessively.

Amy's portable therapist helps with a major transition. From an early age, Amy, now seventeen, has suffered from a severe anxiety disorder that continues now also as separation anxiety as she anticipates leaving for college. In preparation, Amy is planning to take photos of her family on her cell phone. She also requests that I record a brief video for her MP3 player to assure her of safety and help when she needed it.

Comment: Experienced therapists are familiar with the occasional patient who dials their telephone number just to hear the voice message for reassurance. A similar instance is the child to whom they give a small gift as a talisman, transitional object, or souvenir that recalls the therapist and the therapeutic space. Amy too wants to tap the power of interactive media play to shrink the psychological distance she fears so much.

Mike was going through a shy stage, but still craved companionship. During one therapy session of online interactive play, thirteen-year-old Mike reveals spontaneously how he often prefers online text messaging and chat room postings to other ways of interacting with his peers. He says that he feels too exposed in face-to-face encounters. A real time interaction via a Webcam is also too revealing, and the telephone presents painful, awkward silences. Instant messaging is quicker that the slower e-mail. Like many of his age-mates, he prefers instant messaging because it is quick, always available, and does not embarrass with awkward silences.

Mike has discovered that when he is in a depressed mood, he is not as well accepted by peers. But with instant messaging he can appear upbeat to his friends even though he may feel depressed. He can put on a more self-confident and cheerful "face" without revealing his self-consciousness and anxiety through voice or appearance. It also allows him to imagine the other person's face as accepting and pleasant. Mike likes to respond quickly and with wit, and instant messages give his friends time to "get it." He also likes the time he has to think before answering. He likes the excitement he feels at receiving a text message and discovering that a friend is online and making the connection. He likes the group of friends who congregate on line.

Comment: Mike's choice of communication mode matches his current needs more than any other. Mike has a wider range of ways to communicate than kids ever had. An adolescent's preferences can reveal his current level of social comfort in general, or specifically with another individual.

Adam gives an aesthetic critique when he takes me on a musical tour of recent video game soundtracks. Fifteen-year-old Adam suffers from dyslexia and had shown little interest in music before. He compares and contrasts the styles, instrumentations, and compositions of scores and provides an excellent and informative history of influences from earlier games. For example, music in a current video game is heavily influenced by music from a game popular in the 1980s.

Having made this exciting discovery, I encourage his parents to show interest in Adam's budding interest. Later, Adam makes a school project "History of Video Game Music." He receives many compliments from his classmates and teacher.

Comment: Personal taste and more refined aesthetics develop as part of the adolescent's identity. This is a positive example of a boy's talent for venturing on his own into interactive media Education Enrichment Growth Opportunities. With curiosity and eagerness to learn from him, I affirmed the importance of this discovery and alerted his parents to this talent. He then had an opportunity to teach others. This adolescent discovered an original way to enrich his esthetic life, and increase his competence, self-esteem, and social success, and contribute to the lives of others.

ADDITIONAL APPLICATIONS

Parents and older youth are increasingly turning to the Internet for information about their health, and in my experience, seem to treat what they see with surprising naïveté. While it is always good to work with a curious and actively-involved patient, much of the online information is plain wrong. To help with this, I forward links to helpful Web sites, such as self-help groups and credible information about diagnosis and treatment. (SEARCH WORDS: aacap, american psychiatric association, american psychological association, nimh.)

Continuing treatment or assisting the referral physician via e-mail with children away from home is a good way to stay in touch and to follow medication's main and side effects. Communication with parents about appointments and other administrative matters is convenient for obvious reasons. For example, parents can ask questions, forward links they discovered, return questionnaires, and report on daily progress as medication is adjusted. All these professional services become part of the patient's record. Once the therapist thinks of it, he will come upon new opportunities to improve his work with interactive media.

Lately, as social media and messaging have proliferated, I have found it necessary to limit their use, including e-mails, For example, I let people know that I am not accessible 24/7 for messaging of any kind, especially about appointments and other administrative matters. I direct people to using the telephone and reaching me during business hours for such matter.

EXAMPLE

Sandy's medication may be causing her side effects. The twenty-five-year-old has had a mood disorder that has incapacitated her and nearly scuttled her academic career since high school. She has been able to complete college and is now in a PhD program at a good university. This was largely made possible, because, once she left for college, we worked well together via e-mail and holiday home visits; her family was supportive, and we eventually found (after lengthy efforts) an effective regimen of medication that included lithium carbonate. I referred Sandy to a variety of doctors and therapists after she left for college, but she prefers to continue her work with me. I am confident of Sandy's honesty and reliability and am comfortable with this arrangement. Here is a slightly edited actual e-mail exchange:

----- Original Message -----
From: Sandy
To: Dr. Schwarz
Sent: Thursday, September 27, 2007 11:25 PM
Subject: Sandy's sleeping problems

Dr.Schwarz,
For quite a few months now I have had a hard time sleeping through the night. Often (on average, five times a night), I wake up very thirsty and needing to go to the bathroom (which I know goes hand in hand). Although the <sleep medicine> helps me to fall asleep it does not help me stay asleep, and this goes for taking .5 mg of <minor tranquilizer> too. Since I am not sleeping so well, I am really tired during the day and for some reason stay up pretty late at night (about one am). Also, I have a very difficult time waking up and sleep pretty late. Is there any way to fix this, or do I just need to learn to live with it? All in all it is frustrating.

Thanks

Sandy

-----Original Message-----

From: E D Schwarz MD
To: Sandy
Sent: Fri, 28 Sep 2007 10:02 am
Subject: Re: Sandy's sleeping problems

Hi Sandy,

I will look up your record when I am in the office tomorrow and will get back to you.
Regards and take care of your self,
EITAN D SCHWARZ MD
In compliance with the Health Portability and Accountability Act "HIPAA" (rule 104-91), this message is intended only for use of the individual or entity to which it is addressed and may contain information that is privileged, confidential and exempt from disclosure under applicable law. If the reader of this electronic message is not the intended recipient or the employee or agent responsible for delivering the message, you are hereby notified that any dissemination, distribution or copying of this communication is strictly prohibited. If you have received this communication in error or it was forwarded to you without permission from Dr. Schwarz, please forward this message back to the sender at the e-mail address above, delete this message from all mailboxes and any other electronic storage medium and destroy all copies.

----- Original Message -----

From: Sandy
To: Dr. Schwarz
Sent: Friday, September 28, 2007 9:22 AM
Subject: Re: Sandy's sleeping problems

Hey,
There is also something else I need you to do for me. They only gave me half a month of <sleep medicine> because of insurance. So, my Dad called and they said that all they need is a memo from you saying that I can have thirty. Just e-mail it to me and then I will get it to my dad.

Thanks,
Sandy

-----Original Message-----

From: E D SCHWARZ MD
To: Sandy
Sent: Sat, 29 Sep 2007 9:50 am
Subject: Re: Sandy's sleeping problems

Dear Sandy,

Reviewing your record, I see we discontinued your <certain meds> in June-July—I wonder if your sleep problem started then. You are now on <other meds>, and occasional <other meds> 2mg.

Please answer ALL these questions fully:
Do you have any uncomfortable leg movements and mess up your sheets?
Are you snoring or do you stop breathing briefly when you are asleep? Ask your roommates to look into your room.
Do you have a lot of thirst and pee a lot? Don't drink anything after eight pm.
Are you generally more anxious or depressed?
Do you take any form of caffeine after four pm? Don't.
Are you drinking or using any drugs? Don't.
Are you taking naps during the day? Don't.
How are you doing generally and how is your stress level?

What to do now? You have been forgetful, so I would not increase the <med>, even at bedtime. You can try to double up on the <sleeping med> to 4 mg (see attached letter FOR INSURANCE you requested.)

My best
Dr S

----- Original Message -----
From: Sandy
To: Dr. Schwarz
Sent: Sunday, September 30, 2007 7:03 PM
Subject: Re: Sandy's sleeping problems

Do you have any uncomfortable leg movements and mess up your sheets? I will have to look into that.

Are you snoring or do you stop breathing briefly when you are asleep? Ask your roommates to look into your room. I KNOW THAT I SNORE. I WAS TESTED FOR SLEEP APNEA TWO YEARS AGO AND I DIDN'T HAVE IT.

Do you have a lot of thirst and pee a lot? Don't drink anything after eight pm. THIS IS MY BIGGEST PROBLEM! I WAKE UP DEHYDRATED AND THEN DRINK SOMETHING AND THEN PEE ON AVERAGE OF 4-5 TIMES A NIGHT.

Are you generally more anxious or depressed? NOT REALLY.

Do you take any form of caffeine after four pm? Don't. I DON'T DRINK CAFFEINE.

Are you drinking or using any drugs? Don't. I DRINK ON THE WEEKENDS AND NEVER DO DRUGS.

Are you taking naps during the day? Don't. SOMETIMES, BECAUSE I DON'T SLEEP WELL DURING THE NIGHT.

How are you doing generally and how is your stress level? I DON'T HAVE MUCH STRESS, BUT OF COURSE I STILL FIND THINGS TO DWELL ON BECAUSE THAT IS JUST ME.

----- Original Message -----
From: E D Schwarz MD
To: Sandy
Sent: Monday, October 01, 2007 2:59 AM
Subject: Re: Sandy's sleeping problems

Hi Sandy,

I am glad you are generally doing well. This is what I am now thinking:

Among other things, long-term lithium's effect on the kidneys can often cause this type of picture. Here is a link to help you understand fully what might be going on: http://www.medscape.com/viewarticle/417833_7. To access the article, click on this Web address, or cut and paste it into a browser window. That is one reason we have been testing your blood and urine regularly.

The next steps would be to have a fuller evaluation by your internist (or someone he refers you to) to rule out other causes. You can have it in SF. Please arrange it thru/by your health service or referral from your doctor here to an internist or nephrologist. Eventually you might take some meds for it, depending on the cause, but not before a full evaluation. (Use the attached request.)

In the meantime, please get a Li level (as usual, after skipping a dose) now. We should also know your serum electrolytes, BUN, and creatinine and urinalysis to get an idea of how things are. It is important that we make sure your sodium is not too low (attached is the lab requisition.)

Right now, try to go as thirsty as you can at night and sleep with your head as raised as possible.

We can also try to carefully reduce or discontinue the lithium, but we have to carefully weigh that against its importance in helping your mood. If we knew the cause definitely, it might help us decide exactly if that might help, so please get the evaluation by your internist ASAP.

Please take the request to the student health service or where you usually have the tests. Also, please send it to your internist—or I could if you get me his fax or other number.

Regards,
EITAN D SCHWARZ MD

Comment: This actual (identity altered) transcript illustrates a successful therapeutic collaboration established a decade before with an adolescent that has continued via e-mail and occasional visits. In this case, Sandy seems to have developed diabetes insipidus and needs further investigation. She took the initiative to identify a problem, contacted me, and cooperated as I tried to figure out what to do next. She is reliable, and I am confident she will follow through.

Chapter 13: The Future II: FITGOALS™

In the previous chapters, I attempt to enable today's parents to take charge of interactive media and assure thoughtful and constructive use of technology in their homes, as well as starting kids off in the right direction, using these technologies. Looking forward to the future, as our inventiveness brings more new interactive media applications, we must develop new ways to shape their beneficial use.

In this chapter, I show how the Media Plan developed in this book can be implemented on a community or wider basis as a comprehensive system of online services that assures parental oversight and maximize Growth Opportunities. I also intend to create interest in the media and software industries in opening up a new space that integrates neuroscience, children, families, and technology.

We can now visualize a future when you would have a wide range of options the actual physical configuration of media in your home. Furniture, hardware, and software packages designed will be offered to parents to enable better family life and child-development.

Please fast forward. Imagine sitting next to your school-age child; both of you seeing and hearing the same output and but each is making your individual input a keyboard and mouse. The machine is the table itself with an interactive top, or three LCD displays are arranged in a triangular configuration so that this output is easily visible all around, or better yet, you are all interacting around a three-dimensional hologram. Webcams and several wireless input devices, such as keyboards, mice, and game controllers, allow clutter-free use of the equipment.

The family computer space can now be expanded with a larger display and as many input devices as users for family members and visitors. The whole family can join around the display, each member with her own joystick, mouse, or keyboard, exploring and visiting distant sites together, or interacting with another family similarly situated in another part of the world via voice and Webcam technology.

Let's visualize the possibilities five years from now: Greater portability, bio-integration, processing power, and miniaturization; and increasingly innovative human interface devices, personal area networks, and a wide array of input, monitoring, messaging, processing, and output devices not even imaginable in 2010. Homes are now designed with special spaces for family computing to accommodate tasteful furniture and attractive multi-user hardware.

Since about 2010, when online service providers provided increasingly sophisticated age-sensitive systems for parents to filter and time kids' online experiences, a whole new field has been emerging in response to growing demands by parents—Family Information Technology (FIT). The need to focus the uses of interactive media on children and in a comprehensive and integrated manner has finally become obvious. Like Information technology (IT), which has been supporting business activities for over forty years, FIT has developed more recently to support child development and family life.

By 2012, in addition to the original software and services supporting family financial, bill paying, and banking tasks, FIT now also provides integrated support services for health, education, civic, and other family functions. Software and hardware developers and merchants have been competing in the very active FIT markets. One health function supported by FIT is meal preparation for the family—diet and Weekly Menu integrated with the family's demographics and each member's preferences, health needs, dietary guidelines, and currently available fresh produce and sale items in the food market. Diet plans are individualized to each family member.

FITGOALS™: FANCIFUL FANTASIES OR PLAUSIBLE PREDICTIONS? DECIDE FOR YOURSELF.

By 2014, FIT offers much more to assist parents in raising children. Digital electronic technology is now helping parents in other important way, making the job of guiding their children's interactive media use by automating and streamlining the Media Plan easier and more complete. Day-to-day implementation of meal planning is now easier. A growing function offered by FIT has been available and improving as FITGOALS™ (Family Information Technology Growth Opportunity and Learning Service) for some years.

FITGOALS™ matches children's and family's psychological and emotional health needs with a selection of the most appropriate and best variety of hardware, software, and Growth Opportunities.

The trend that began with incorporating the Internet, video games, DVDs and videos, interactive television, intelligent toys, mobile phones, and MP3 players in smaller devices continued. Devices can be connected through wireless technologies, and thereby all media are part of an integrated system that can also be connected to a point of purchase monitor and other inputs. Cloud computing is the movement of software applications and services from COMPUTERs to centralized data centers, where they are made available via the Internet.

FITGOALS™ systems are but one likely consequence of the growth of cloud computing and integration of many digital functions into wireless devices. The service is the only way a child accesses any interactive device, including cell phones, Internet links, and even TV. Such systems make it possible for parents to create and monitor Media Plans. A child has only one way to enter the system—through the parent-designed FITGOALS™ portal.

FITGOALS™ systems are designed from the ground up to support families and children. They are essentially gated digital communities to promote family support and healthy child development in the use of all interactive devices, including telephones, handhelds, and all new gadgets. The incorporation of input, processing and storage, and output functions of various applications within one device, together with the interconnectivity of devices through wireless technologies, enables FITGOALS™ to encompass every media device and digital application available to children. The service acts as the sole portal through which the child can access all interactive media activities. It is also an elaborate system of databases and search engines that serves parents in designing Media Plans for each child.

FITGOALS™ provides each child his own easy to use digital media "dining room" with his own unique daily menu within the constraints and choices set up earlier by his parents, and then serves him his interactive media "meal" by linking him to preplanned applications, monitoring his usage, and moving him on. FITGOALS™ would be designed by parents for their individual child with as much flexibility as they desire, guided by the methods described in this book.

Online and interactive media menus for children with special needs are based on the latest scientific data. On a regular basis, as conditions change, the family modifies the profiles in their database, and the FITGOALS™ adjusts its guidelines and menus. FITGOALS™ empower parental control and teach responsibility and good judgment.

By cataloging, classifying, and matching Internet and other digital technology content with extensive search engines and review by editorial staffs, FITGOALS™ can provide content designed to enhance brain, neurocognitive, and emotional development, and psychological health for specific age groups. Building on discoveries made and techniques developed over the past half century, online modules are individually tailored to prevent and remediate deficits or delays in functions like attention, memory, and reading and math skills with game-like exercises in normal or challenged children. Other programs detect, prevent, or delay onset of certain mood or anxiety disorders in vulnerable kids. Up-to-date, integrated, comprehensive modules for other special needs children supplement person-to-person remediation by parents and professionals.

To guide and automate the tasks of creating healthy Media Plans and implementing menus for each child, parents select a FITGOALS™ subscription from among several available to consumers. In selecting a FITGOALS™, they look for ease of use, comprehensiveness, professional and psychological soundness, family friendliness, user trust, and a good fit with their lifestyle and values. They have an easier time than their own parents did in putting it all together and providing their children with the correct mix of Growth Opportunities.

FITGOALS™ educates and guides subscribers pretty much as this book had done for their parents in providing a Media Plan. They begin by inputting the family's demographic and preference profiles into a secure questionnaire akin to those still used by online dating services. (FITGOALS™ user privacy has been protected by law and guaranteed by active technical safeguards since 2015.) Along the way, the service teaches them the latest about child and family development, much as their parents learned about paying taxes using home financial software.

Many FITGOALS™ systems are huge media companies that provide most technical and content services in house. Some small FITGOALS™ systems are staffed by editors, scholarly researchers, and technicians focusing on services to special groups, like the hearing impaired community. A specialized FITGOALS™ service also licenses signing avatar systems, captioning, and other specialized modules to other FITGOALS™ systems that want to provide access to the hearing impaired. Some entities are identified as brands, but actually have only a small staff that supervises a gateway to what appears to be a seamless affiliation of specialized franchises run by a variety of site operators all over the world.

The history of these developments is interesting. The FITGOALS™ industry evolved rapidly, typical of the information technology field, appealing to large sectors of the family and educational markets. The services are structured in a variety of ways but all have family and child-friendly content in common. The first FITGOALS™ was created and operated with capital investments in the private sector beginning in 2010.

By 2012, the FITGOALS™ system was in place as a large and growing sector of the media industry, combining free enterprise and government regulation. Additionally, non-government organizations like the Boy Scouts, Parent-Teacher Associations, health, education, and legal professional organizations, and special interest and consumer groups were providing ongoing scrutiny and recommendations. Some jurisdictions were even providing access to a FITGOALS™ to their residents free of charge, much as they provide other services.

Because they had become services essential for the public good, it soon became clear they were necessary for the public welfare because they supported family life and healthy child development. Yet standards and practices varied too widely. After vigorous political debate, passage of a number of federal and state laws, and testing in court cases, a well-functioning system of regulated, profitable, private FITGOALS™ utilities evolved, and was integrated with systems in other countries as the Internet was.

How Parents Use FITGOALS™. Gone are the days of labor-intensive Media Plan designing and online and interactive media menu design, haphazard monitoring, power conflicts, and Internet and other media perils. With FITGOALS™, parents assure that each child has a plan she needs for her development delivered in a mix of the most palatable Growth Opportunities available. Parents now easily plan interactive media menus that provide the best mix of essential online and interactive media nutrients to fit each child's needs, consistent with the family's values and child-rearing practices.

FITGOALS™ initially walks parents smoothly and flexibly through the entire process presented in this book supplemented by search and editorial services. FITGOALS™ samples the entire Internet, categorizing, rating, and filtering content to present a limited number of selected choices suitable for families and individuals, maintaining a high degree of user trust and confidence. For example, FITGOALS™ offers only unbiased and reliable medical or legal information, shunning the myriad commercial and testimonial-driven sites that do not have the interest of the consumer in mind.

These services also able to link the children's school and health records into the family's data FITGOALS™ base. They can consult specialized libraries and online experts. All this information is processed and integrated by FITGOALS™ software that utilizes vast, well-designed, constantly updated databases on child development, the needs of families and special needs children, available hardware, and local and distant Web sites and services, as well as "on the ground" community resources and services.

Parents in 2015 use FITGOALS™ as a safe child- and family-friendly portal to the Internet and all other digital technologies kids use. With a level of flexibility or strictness they determine ahead of time, the FITGOALS™ creates, monitors, tracks, and manages Media Plans and menus in real time for each child. As a matter of fact, a law passed ten years ago requires children younger than eighteen to access the Internet and other digital technologies only through FITGOALS™ portal approved and designed by their parents.

Its effects on family life and child development are constantly monitored by FITGOALS™ and adjusted to maximize its benefits. The parent-driven FITGOALS™ system has grown in a logical and rational way to make parenting both more effective and easier.

The Future II: FITGOALS™

FITGOALS
Family Information Technology Growth Opportunities and Learning System
Guides parent to customize media diet for a child and then provides daily media menus to the child through licensed portals
www.mydigitalfamily.org

An innovative media system that supports family life and healthy youth development and encourages visionary licensed software, hardware and content providers and marketers with incentives to deliver pro-social media alternatives into popular culture

- Benefits family life and psychosocial development of children 0-18+
- Guides parents to seamlessly select, design, monitor, customize and control a child's balanced media diet
- Encourages a balanced media diet of developmental opportunities: family relationships, values education, socialization, education enrichment, and entertainment
- Allocates developmental opportunities and schedules media activities, quantifies and schedules parent attendance, presents daily media Menus and actual meals, and guides child's usage
- Monitors and limits the child to healthy and safe licensed media and device usage and guides parents with feedback and recommendations
- Comprehensive, flexible, age-dependent multisite, multiplatform, integrated and interactive
- THE GOOD LIFE is a pro-social multiplayer action video game that integrates with FITGOALS
- Offers potential for wide acceptance and vast social good

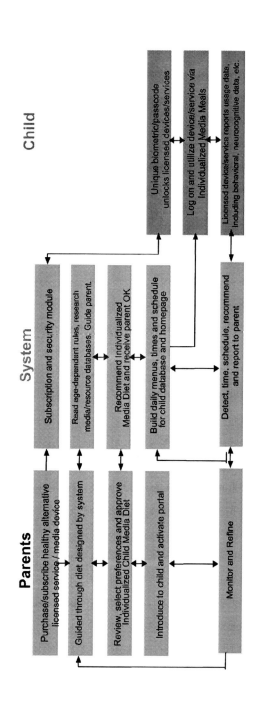

EXAMPLE

Jane is now married—to Dick, of course (they met online)—and they have three children and a home of their own. Both were raised with good Media Plans by parents who labored to provide their children with a casual, and sometimes gourmet version, made of the best balance of essential online and interactive media nutrients found in the best mix. Each time Dick and Jane remember how hard their parents worked at this task, they feel a wave of gratitude and respect. Naturally, they are familiar with interactive media that has developed in ways unimaginable twenty years before. Like most couples, they remodeled a room in their older home as a nice media space to enable family computing and roaming online to accommodate the latest interactive media. It is now easier for them to make sure that the children are subject to the same system of privileges and limits set for all other areas of their lives.

Let's focus on seven-year-old Gena. Since she was a toddler, FITGOALS™ has assisted Jane to find the right balance of Growth Opportunities to promote Gena's development. Dick and Jane have deliberately established themselves as an authentic presence in her life to provide her with the scaffolding she has needed at every phase. FITGOALS™ has guided them to mentor her into collaborating as their partner in this effort as she has matured. Her parents have not been aiming at perfection—from themselves or her.

They have given her enough room to take initiative. They expect her to make mistakes that show she is stretching herself and to learn how to fail as a basis for a solid belief that she can successfully impact her world, know what she is good at, not to give up easily, and to enjoy what does. They have coached her to act in ways that give her a sense of competent mastery while also remaining generous, fair, and inclusive of others.

They have been mindful to help her retain her own thinking and judgment, especially in their own relationship with her. They have taught her to be vigilant about the power of interactive media. For example, her mother sets the "independence function" so that seven-year-old Gena can override the FITGOALS™ restrictions and roam online on her own, visiting filtered sites of her own choosing.

If Gena wants to stay longer at a site, she can override the presets, but she must type in a reason that will later appear on the parental log for review and discussion. The service quickly readjusts and recalculates an alternative menu for the rest of the week, giving Gena choices among several options.

In addition, her parents have consented to allow Gena's teachers to add interactive and online activities apart from their own menu. Teachers can access the FITGOALS™ of all students in Gena's class, all at the same time, to provide curricula. Sometimes the FITGOALS™ itself comes up with innovative menu suggestions. This automated, intelligent, and flexible service frees up parent time and removes them from the role of vigilant enforcers of family online rules. Additionally, it further reduces the number of occasions for conflict.

Jane uses the FITGOALS™ to find like-minded moms, some even residing in other countries. Jane is trying to talk Dick into becoming more active and joining online communities of families that advocate for family-friendly international causes through their FITGOALS™.

Today, Gena sits at her family computer and goes online. Gena's parents are, as are most American parents now, excited about their child's interactive media use, and are confident that it brings great benefits at almost no risk. All children now gain access to interactive media at home, school, or mobile, only through a media portal customized with an individual Interactive family media plan. Gena and her parents created her plan with a streamlined step-by-step procedure online at a site provided by FITGOALS™.

First, Gena provides identification (biological "password" or thumb print, voice, eye, or facial recognition), and activates her own security fob with a biological metric which then becomes active for several seconds. The key works something like the old-fashioned 2008 automobile keys or credit cards, containing unique programmed digital code. It has become fashionable to wear the fob embedded into her ring, belt buckle, or other piece of jewelry. She passes the activated key near the computer or mobile device sensor. The sensor is also coded and activates several layers of verification and security. The uses of this system are limited by law to maintain privacy, safety, and civil rights.

Gena's menu for the day appears and offers her the schedule of preselected sites. Effortlessly, the FITGOALS™ service links Gena to her selection and times and monitors her stay at each. At the proper time, the software moves her through the menu seamlessly from item to item in a friendly, but mostly undefeatable way.

Because she has shown good judgment and responsibility, Gena is additionally allowed some totally unrestricted online time. As Gena gets older, she will have more and more independent choices and privacy. She sometimes flags sites that would provide Growth Opportunities for other family members as well, for addition to the recommended list that is regularly reviewed by her parents.

Using their own log-on procedures, other users are permitted by her parents to join Gena to widen Socialization and Family Relationships Growth Opportunities. These may include family members, teachers, and friends. The FITGOALS™ occasionally suggests sites based on Gena's interests. But when she wanders into sites the FITGOALS™ had designated as unapproved, a warning reminder of the previously agreed-upon consequences appears, giving Gena a choice of staying or leaving the site. This arrangement allows for teaching freedom with responsibility.

Similarly, Gena's cell phone calls are routed through her FITGOALS™ portal. The FITGOALS™ can report to her parents how well Gena's Media Plan is working and suggest modifications.

Chapter 14: The Future III: A Nonviolent Video game

Play is so important that most young mammals come prewired to initiate and evolve their own play. Such play has deep biological evolutionary roots. They seem to enjoy emulating adults and learning, refining, and practicing species-specific physical, social, and survival skills that go beyond the range of everyday life.

And their elders instinctively safeguard their young as they allow play. In a similar way, human play reflects and, in turn, influences our species and cultures. Human children's play reflects how they emulate adults and experiment with social roles and behaviors. Often, they adopt heroes that embody what they admire and who they seek to become.

Technological toys have a way of working themselves down into younger and younger age groups. Let us not forget that video games were originally invented in 1958 by an American engineer looking for a way to have fun, with his intended audiences of adults. So when we survey the video game landscape now, we are looking at much of what very young children will have in their cradles in some form within a short time.

While I agree that games have a seamy underbelly, as do magazines and films, they do inevitably reflect our culture back to us. They can also be a powerful and positive force in shaping our culture. I believe that we can influence children's development by using these powerful technologies.

In this chapter, I build on prior understanding of the role of children's play in their development. I review the history of video games with an eye toward age compression. I then let my imagination run loose to show how enormously beneficial an action video game might actually be to children.

A type of game that I believe could be extremely helpful to children has not yet been marketed, although the technology is all there. It can be nonviolent and engaging, and at the same time assist children to become responsible and productive. It can simulate the real-life choices that children face as they develop. Unlike current games, such a game would put a player "on the ground" as a person making choices in his own daily life, test out scenarios, and live with the consequences of his actions. I demonstrate how a game can promote individual responsibility, loyalty to family, community service, and internationalism. It can reward players for good deeds, kindness, and forgiveness. And it can vigorously promote innovative Education Enrichment, Family Relationships, and Socialization, as well as Values Education Growth Opportunities for children of all ages.

The Future III: A Nonviolent Video game

HISTORY

Over the decades, with the evolution of consumer digital technology and the addition of the joystick and other human interface devices, video games have been working themselves deeper into early childhood. They reflect, and in turn create, popular culture in the United States and Japan. Early in the history of video games, competitive but nonviolent games like Pong and Pac Man, intended for both genders and most age groups, became great commercial successes that transformed the home TV into an interactive medium. In the middle 1980s, Japan's Nintendo successfully released its Famicon system (FAMily CONsole) to children and families. In Japan, a strong traditional family is a dominant institution. But in the United States, where families have become less important, a reference to family was considered harmful to the sales of Entertainment systems. So in the United States, Nintendo's Super Mario Brothers boomed without reference to the family. That tells us something we already know about the decline of family life in our culture.

Limited by scant computing power, early characters and their movements were simple, cute, and childish, and sound tracks were made up of simple tones and beeps. Gaming grew, along with digital electronics, by offering more elaborate and interesting characters, actions, stories, and sound tracks. Play with action figures like G.I. Joe had long served as acceptable doll play for boys. Video game characters took on increasingly human forms. However, unlike play with actual dolls, early games were rigid and determined the nature of the player's experience, leaving little room for a player's imagination, initiative, creativity and inventiveness.

Later, games started to invite players to design or modify the game. In the past decade, the gaming experience has evolved in complex and innovative ways. A gamer—or community of gamers—is encouraged by the manufacturer to alter and customize characters, actions, and rules. Gamers modify ("mod") the form and content of the game itself, generating a different game or even making an animated short story—"machinema." In this way, video games allow players to exercise imagination, creativity, and technical skills to a much greater extent than before.

According to some concerned people, children, especially boys, are spending too much time playing such violent video games that have no redeeming value. Since the 1990s, advances in technology have spurred violent and downright nasty video games. Three-dimensional, first-person shooter games like the early Wolfenstein 3D and Doom, which paralleled cynical alienated music trends, sparked congressional clamor, and the gaming industry developed a rating system in response.

Today, few American young adults have known a world without digital gaming. Since the 1990s Internet explosion, MMOs (Massively Multiplayer Online fantasy games) have drawn millions of adults and youngsters the world over into challenging interactions, even including professional competition and international conferences. Increasingly complex games are now catering to millions of sophisticated gamers and their children in ways that challenge the boundary between reality and virtual reality.

More recently, MMOs have evolved to include large social and moral dimensions even more reminiscent of doll play. Unlike the rigid structure of earlier games, a game's artificial intelligence offers and responds to the player's specific choices, making each player's game unique. In a game like Black and White, the player is empowered as a demigod to choose his own personality and make his own moral decisions about other creatures. Intelligent games like Sim City and Civilization give players tools to create their own design and build and operate a virtual world.

Alternative worlds where players live alternative lives are more lifelike MMOs. Players create their own dolls or avatars, human, inanimate, or animal, and dress them up in unique "skins" to interact with other avatars socially and economically. The line between game and reality is blurred as these parallel universes engage more aspects of players' emotional lives. These experiences are extremely meaningful for many, who sometimes also actually go on to meet in person. Entropia and Second Life offer social and economic platforms with millions of players. These virtual worlds engage players in virtual economies where actual real-world money is exchanged and properties and items are bought and sold. Actual businesses have opened virtual stores to sell virtual products for real-world money for use by game avatars. There are individuals who actually earn substantial incomes in this manner.

Paradoxically, the American military, without which video game technologies could not have been initially developed, has turned to the video game industry to create training simulations. America's Army is available for free download online. Such games assist the military in its efforts to recruit new soldiers. Weapon systems are now designed to utilize the skills they have developed over years of playing video games. Kids are already well practiced in pushing a button to zap an enemy. Worldwide, game-like simulations have become primary means of military training and rehearsing war scenarios. The line between Entertainment and actual military reality is likely to become blurrier as the military and gaming industry collaborate more closely. Some critics worry that this trend encourages love of war, and point out that games are games and war is war. Games present war as a sanitized experience, where actions have few real consequences, while a real war has horror, ugliness, human suffering, and death.

Sports games allow players to move characters realistically and enjoy mastery of games and their rules. In fact, virtual reality is helpful in many applications. Currently it is used for learning negotiation, tactics, and the treatment of posttraumatic stress disorder. Compassionate games like Darfur is Dying have players experience powerlessness and compassion. Games for Change, or G4C, supports gamers and designers of games with strong pro-social content. PeaceMaker simulates making peace between political groups and nations.

I am sure that many of these historical trends will reach younger and younger children, who will never know a time when they did not play with intelligent machines. What follows below is a magazine feature that might appear in about five years.

THE GOOD LIFE

It is now 2015. Just as junk foods still abound, popular culture still lures children with vulgar and exploitative media contents. However, commercially successful games such as *The Good Life* have evolved as a result of the FITGOALS™ industry and partnerships among devoted grass roots groups and nonprofits with commercial entities to provide alternative healthier choices to children and parents.

In today's *The Good Life* session, Steve and Nathan are continuing their defense forces reserve duty. Defending the community is a Good. They are collaborating in planning and operationalizing an effort to unmask an avatar they suspect to be an enemy Ungood. They challenge the suspect with a variety of dexterity and brain teaser contests which they obtain from the game's arsenal. No violence is allowed. But the avatar turns out to be a legitimate citizen, and although they apologize profusely, they are both penalized Goodness point values for mistreating a fellow citizen. Every action, no matter how intended, has its consequences.

Unfortunately, this penalty occurred when actual point values are at an unusual peak. Luckily for Nathan, he had already spent most of his points on Goodness power, and the remainder is stowed safely in the Community Goodness Fund. Saving is good. But Henry's Goodness point wealth plummets. Steve has been careless by not buying Goodness power with his points and not saving in the Community Goodness Fund. Because the particularly large exchange rate is relatively rare and the likelihood of making up for it relatively unlikely, Steve suffers heavy losses that will require a long time to regain. Sometimes life is unfair.

Steve is dejected and wants to quit, but Nathan talks him into continuing. Their friendship is good, and Steve stays. Henry takes on a community project together with a player from China he had not met before. He cleans a toxic waste dump and is able to earn some extra Goodness points. Community service is good. Internationalism is good.

Next, Steve does game schoolwork, continuing to learn the information and skills required for graduation to the next level. He is working his way to becoming a reporter, and today reads historical writings that underlie modern journalism. He passes a quiz on the subject and earns points. Self improvement is good.

But Steve has been neglecting his game family, and the penalty for this offense lessens his wealth even more. In the meantime, Nathan is at home with his game family, also earning points for his Good Deeds and depositing them quickly in the Community Goodness Fund. Family is good.

Nathan's interactive family media plan, incorporated into his FITGOALS™ plan, includes special needs Education Enrichment for a child with attention problems. FITGOALS™ had been customized by his parents to progressively train him to ignore distractions and pay attention better. He did well that day, and he earned extra points. Working on your problems is good. His parents will see the log later.

Other challenges will come up next time, including a forest fire that requires all citizens to pitch in with physical challenges.

The Good Life was actually first developed by a small independent company. It became a FITGOALS™ module in 2015, integrated seamlessly with its other FITGOALS™ components. In the next ten years, modules created by other small companies have been added to *The Good Life* under the editorial control of its FITGOALS™ and own board of directors. Standalone or handheld versions can be played online and must often be updated and authenticated with online updates.

Although initially marketed as a much less elaborate basic version and modest commercial effort in 2010, the game has become a huge commercial success. *The Good Life* is now supported by foundations and businesses as a public service in some countries, depending on its integration with FITGOALS™. A large corporate-like staff, supported by an international cadre of consultants, provides technical and content development, marketing, growth, and operations. Like all games, *The Good Life* is accessible only through FITGOALS™. In fact, children whose families have not yet adopted FITGOALS™ often nag parents so that they can play the game.

The staff also safeguards players' safety and privacy and oversees daily operations to prevent attempts to subvert, hijack, or misuse the game for anything other than its chartered purpose. *The Good Life* entity is governed by a distinguished international board made up of representatives from national boards (similar to the Olympics, and with strict ethical rules) that may be made up of boards from other jurisdictions to assure its mission and success. Boards include clergy, business, government, and professional leaders, FITGOALS™ CEOs, and Nobel laureates. The most successful three players also sit on its board.

Well positioned competitively with other popular games, *The Good Life* can take advantage of the full range of the latest interactive media, human interface devices, wireless personal area networks, and central processing to transport the player into a virtual play reality abundant with magical illusions.

For example, in an advanced form of the game, the player, represented by her avatar, moves within real-time holograms or other types of virtual reality spaces, much like the 3-D Holodeck of *Star Trek*. Multi-channeled surround sound with special effects provides her with an auditory virtual space. Through a variety of devices, she can feel multi-sensory computer output effects like temperature, wind, touch, pressure, and smells. Vibrations or movements of her chair or the floor add to the reality of her illusory experiences. She can input text, but communication is usually through real-time spoken words altered for her avatar to add interest and mask her identity. When necessary, her language is translated in real time to other languages or signed by a signing avatar.

Input sensors in a mat on the floor and motion detectors enter her movement and location. Other input sensors detect her posture, position, facial expression, pulse, blood pressure, respiration, and skin conductance. Powerful online computers, supported by ample broadband links, receive and process input and almost instantaneously send output so that players have the illusion of interacting together in real time.

The Good Life has the same basic structure that has been common to massive multiplayer on (and off) line role playing action video games (MMORPGs), already popular for decades. It offers conflict between good and bad guys, great graphics, plenty of action, competition, challenge, and suspense. But *The Good Life* differs from all games Dick played when he was younger—it is aimed primarily at teaching and practicing values, personal and social responsibility, and building leadership skills. It contains no violence.

Players are rewarded for performing Good Deeds with minimum of presuppositions that they will necessarily benefit materially or that life is always fair. Good Deeds are socially valuable and morally correct acts that contribute to the family, community, society and the common good, and are cherished by all civilized people. These include acts of charity, generosity, virtue, creativity, community service, and defense of human rights and freedom.

The game encourages a diversity and balance among Good Deeds: self-improvement; consistent persistence toward a goal; family or community loyalty with a generous helping of internationalism; freedom, spontaneity, entrepreneurship or initiative with responsibility; fairness, justice, or self-interest with forgiveness, mercy, and compassion; defensive war with peace; and problem solving, especially with international partners, with individual action. The game punishes actions opposite Good Deeds with point loss, but appeal is possible.

Action and conflict are generously sprinkled throughout to keep players engaged and to keep *The Good Life* a challenging and fun game. For example, the game itself generates a never ending variety of exotic and clever creatures that threaten players or their societies. *The Good Life* moves fast and requires dexterity and good eye-hand coordination.

Players advance through hierarchies of levels and sublevels by earning Goodness points and using these to buy strengths or abilities, subject to complex social rules and to chance. A player is rewarded for following in the productive path of an idealized hero. Points accrue for actions that contribute to the common good: taking care of family, community service, defending the community, performing surgical procedures, or rescuing a neighbor or stranger. The game has its own specialized terms, which include Good Deeds, Misdeeds, exchange rate, Goodness Power, Stature, hero, path, mark, league, transform, defender, acronyms, and neologisms.

The Good Life has spawned an international community, akin in some ways to the earlier SecondLife.com. People of all ages in many countries play it on- or offline over long periods of time. In fact, a player displays his own national marker when he enters an international game and is rewarded for partnering with international players. A player can enter any game anywhere. Players from many countries join to compete and learn about values common to all people and thereby learn tolerance and respect for each other. It provides real-time translation that enables international play and chatting.

On a regular basis, *The Good Life* posts summaries of rates of Good Deeds made and received by the players of each state and country to encourage players' hopes for peace among nations. Public media throughout the world keep track of this competition as sports news. The game requires many decisions that reveal a player's character, thereby allowing virtual friendships to develop.

The technology adapts to each player. The game allows physically or intellectually challenged children to play through specialized human interface devices. Depending on the age and attention capacities of a player, the game assures that does not over-stimulate by fast-moving scenes. Because decision making and problem solving are at least equal to actual outcomes of actions, the game keeps track of each player's reaction time and adjusts its level of difficulty to the player without giving an unfair advantage. It may even provide adjustments or actual advice to improve performance. Players are sometimes surprised by unexpected turns of events. Simple and complex incentives encourage the practice of a wide range of universal values, but randomness and setbacks also play important parts.

FITGOALS™ uses the player's age and profile, including unique limits, special needs, and parent preferences, to automatically set its level of difficulty, other special features, and illusory power to achieve a mix of fun and safety. The game monitors and reports duration and frequency of play and other information enabling parent supervision to the FITGOALS™.

For younger children, the game suits the level of their cognitive, social, and moral development with less complexity, ambiguity, and lesser challenges, and requires a parent's presence. And for older kids and grown-ups, the game has more complex social hierarchies and collaborations and greater ethical and moral challenges and ambiguities. For example, avatar judges and leaders may face complex problems or there may be some enemies who are not totally bad and some friends who are not all good. Questions like, "Does the end justify the means?" arise, and working within shades of grey requires more thoughtful and compassionate decision-making, appropriate to adolescent Values Education.

More advanced levels of *The Good Life* include a heavier, more annotated academic emphasis with references to philosophy and history, greater amounts of didactic material, and quizzes that count as Good Deeds for self-improvement. Like Wikipedia (still a popular resource in YEAR), it has a large knowledge-base created by the public and supervised by experts. Ethicists, theologians, and clergy have also added modules that are used for academic credit in high schools and colleges the world over. A text chat function with real-time translation enables and encourages quality text communications. In this way, a player has an opportunity to improve writing skills and to appreciate the diversity and goodness of people. A community of players posts original poetry and art.

For educators, the game is an important tool for the common good and individual health as a diagnostic, prevention, treatment, and research tool. Because the game requires the player to provide real-time aesthetic, intellectual, and neurocognitive (attention, memory, eye-hand coordination, sequencing, reading speed and comprehension, auditory, visual, and spatial processing, etc.) inputs and reactions to stimuli, the game reports to parents their child's level of maturity. It can compare the child's performance to others in its standardized database and keep track of his advances along his developmental trajectories over the years.

Using this information, parents can also readjust the Interactive family media plan or reconfigure both the game and FITGOALS™ to provide the child with optimal fun and development experiences online and in the game itself. Parents can target Growth Opportunities as well as literacy, arithmetic, language, and other skills.

The game often picks up very early subtle signs of attention, learning, and psychological problems before they show up in school or at home, making it an extremely valuable tool for monitoring children's progress along certain neurocognitive and emotional developmental trajectories. Early detection makes prevention or early remediation possible before problems fester and become major impediments.

The game continues to have all possible safety, privacy protection, and security measures. The FITGOALS™ integrates all such games to minimize hazards. Player authentication is by non-defeatable confidential biometrics or other available non-intrusive fail-safe means. Allowed times of day and daily use times are determined by the child's FITGOALS™ settings. Additionally, the game offers ethical educators, psychologists, and neuroscientists a rich and huge cross-sectional and longitudinal data base to study normal and abnormal brain maturation and development disorders and their prevention and treatment.

Scientists can apply to access the information or even design studies with control groups, but informed consent, anonymity, privacy, and parent permission are strictly required. Applications are reviewed much as they are in hospital review boards, and the studies are additionally monitored by a diverse board for safety and ethics. Additionally, the game can detect hackers and abusive or unusual patterns of online behavior and immediately terminates and blocks reentry without parental permission.

Like other video games in 2015, *The Good Life* can also simulate so many aspects of reality that it can pose risks to some players who may not be able to recognize it fully as an illusion. By virtue of age, developmental, or health reasons, individuals vary in their ability to make the crucial distinction between reality and illusion and make the transition from playing to reality smoothly. While we all temporarily suspend attention to this distinction to enjoy movies, for example, some individuals cannot do so completely. Some people may be haunted by or unable to let go of a compelling illusory image or experience, or are compulsively addicted. Therefore, all games that can create elaborate and powerful illusions are carefully regulated and require close parental monitoring.

Players' firm realization that it is only a game is assured in several ways. For example, younger players are restricted to the on-screen version operated by voice commands or a controller only in the presence of a parent. Similar to a driver's license, a player must have a valid license that defines her specific level of permissible illusion power and level of difficulty. Obtaining a license has several steps.

Players must progress through simple miniature hologram versions to qualify for the life-size holograms. Parents must fine-tune the game's settings and test drive it before a minor is allowed to register. Parental permission is required regularly to assure that the game has no adverse effect on their child. A probationary period of temporary licensing assures a good fit, including parental test driving. Frequent, periodic relicensing is required. To change a level of illusion power, the process must be repeated.

In addition, upon logging off, a player must go through an interactive customized transitional period, also related to the intensity and duration of the session, to assure safe return to ordinary daily reality, much like the slow ascent of a deep-sea diver to prevent the bends. If a player does not comply with the regimen, her next attempt to log on is blocked by the FITGOALS™ service until she obtains parental release. Cheating is difficult because parents, too, must be authenticated by the system.

Steve and Nathan Play. Parents work with their children to customize the level of difficulty and make a safe and fun fit to enable the game to customize its level of difficulty, its special features, and illusory power to give Steve a good mix of Growth Opportunities. FITGOALS™ service recognizes Steve's license and monitors him. For example, the FITGOALS™ service makes sure he plays no more than three hours a week, with no more than forty-five minutes a session.

To prevent solitary play and maintain the Socialization opportunity, his parents have set his license to allow play only together with a relative or known friend in the same room. Dick plays with Steve often, thereby also monitoring and enjoying a father-son experience. Sometimes the whole family plays with other families or long-distance relatives, who live far away.

His parents are pleased to steer Steve toward this activity, since it offers rich Growth Opportunities in many areas, including Family Relationships, Socialization, Education Enrichment and Entertainment, without violence. Steve receives special opportunities for depression prevention, but most of all, *The Good Life* gives Steve practice in the application of values. Jane also requested that the game evaluate and remediate Steve's reading and attention skills.

The Future III: A Nonviolent Video game

Twelve-year-old Steve logs on to *The Good Life* at his friend Nathan's house through his own FITGOALS™ portal. The FITGOALS™ recognizes and calibrates itself to both boys. The children like the challenges of its action, fantasy, and complexity. Because the game is so wholesome, sometimes their parents give them freedom to play outside their Media Plan regimen. Their allotted play time today is ninety minutes.

Nathan's house has a great media room with holograms and vibrating floor pads and great sound, and even some smells, providing high illusory power they both love. Nathan is hearing impaired and requires a signing avatar and captioning, automatically provided by the game.

The boys' game is set in Mobujemoland (an acronym after Moses, Buddha, Jesus, and Mohammad), one Good Land among several within an elaborate fantasy that occupies outer space and underground and undersea locations. Each Good Land has its unique mythology and environment. Human-like user-customizable Citizen avatars inhabit, interact, and advance along life-paths within institutions and social structures parallel to ours.

The game itself generates exotic Rogue Lands, each with its own character and structure, populated by an ever-changing variety of unethical and corrupt leaders and malevolent creatures called Ungoods that often threaten the Good Lands, posing challenges to players. Governments of Good Lands generally strive toward the good, while those of Rogue Lands strive toward the opposite, but the line is sometimes blurry, posing interesting challenges to older players. For instance, Citizens of a Good Land can create and support a rebel force and try to transform the Rogue Land into a Good Land.

The aim of the game is for the player's avatar to achieve the high personal goals the player had set initially. He accomplishes his goals by building a productive, participatory, moral, and ethical life in a Good Land. The player is also challenged by moral and ethical choices, as well as the need for defense against Ungoods in a variety of conflicts. He must take initiative and sufficient risks to fail at times, stretching his abilities and teaching how to recover. The sum total of his progress is signified by his increasing Stature.

The cumulative result of a player's decisions, gains, and setbacks is Stature. Stature is the ultimate and only measure of progress. A player starts out with a minimal amount of Stature to reflect the dignity of all people. Each time a player spends Goodness Power for a Good Deed, he also gains in Stature, regardless of the outcome. Stature also enables a player to start his own family, become a defender, and advance through levels and ranks as an Officer.

Stature is relatively stable—only Misdeeds can lessen a player's Stature by producing Shame. Sufficient serious Misdeeds, or merely the absence of Good Deeds, can rarely result in a morally bankrupt player, who loses his citizenship and turns into an Ungood. A player's Stature is not easily evident, providing a sense of equality.

A player must always retain and maintain her avatar's Stature and identity, no matter when or where she plays—online with all other players, individually, or locally among relatives and friends. Steve has played this game for several years with many partners, including his sister and parents, as well as online with many others. He is in the high end of the lowest level of Stature—school—and will soon be eligible to graduate to the next level.

Progress is subject to challenges, competition, dangers, and luck. A player progresses through a hierarchy of leadership positions by doing Good Deeds. The game rewards Good Deeds according to their generally accepted significance, although it can be adjusted to reflect cultural diversity and is reviewed regularly by the international board. For example, saving a life is highly rewarded; charitable acts can range from a lesser value of putting money directly in a beggar's hand and looking him in the eye, to anonymously enabling an unfortunate person to regain his ability to support himself and his family. Good Deeds that are original are rewarded extremely well, as are those initiated by a player or a community of players.

Other Good Deeds include: Aesthetically, socially, and scientifically creative or innovative works; voluntary participation in a neurocognitive or psycho-educational study; raising children and taking extra time with the family, leading a city Civic Council in good works, catching a criminal, working to preserve the environment; curing a patient; mentoring, teaching, learning; apologizing, rehabilitating prisoners; forgiving; and starting and raising a family and maintaining good marital and parenting relationships.

Good Deeds include the works of talented performers in the arts and athletics who help people appreciate and celebrate the beauty, grace, prowess, and genius of the human body and mind. While individuals must sometimes stand up for what is right, and some Good Deeds must be done by a single individual, joining forces with others whenever possible is also encouraged. Misdeeds are the opposite of Good Deeds, and include shaming, injuring, or stealing from another player, especially the stranger from another country or the infirm; treason; greed; and neglect of family and civic duties.

The program favors direct help to avatars of players from other countries and keeps track of Good Deeds for each player.

Completing a Good Deed earns a player Goodness Points. The more significant the deed, the more Goodness Points she earns. A hierarchy of Goodness Points is published, and earning Goodness Points is the actual immediate objective of players. The faster and more intelligently a player acts, the better he does.

Because intentions are not enough—only actions count, Goodness Points in themselves are without value. Goodness Points must be turned quickly into an ability to act—Goodness Power.

There are no free lunches—doing Good Deeds does take sacrifice. A player must spend Goodness Power from her store to do more Good Deeds (or continue the same Good Deed). She then earns Goodness Points again and the cycle repeats. Generally, a player does come out ahead in net gain.

However, Goodness Points must be used and cannot be hoarded. If not otherwise used to do Good Deeds, Goodness Points and Goodness Power gradually lose value at a random rate, thereby slowing, stopping, or reversing a player's progress. A person cannot rest on his laurels, but must keep doing Good Deeds.

A Good Deed does not always bring rewards. For example, what seemed to initially be a good choice can result in adverse consequences because the player did not choose an alternative. The player must bear responsibility and cost, even though his intentions may be good.

And there is more: randomness is always a factor, and a Good Deed does not necessarily guarantee consistent rewards. Since the actual value of a Goodness Point comes down to how much Goodness Power it can buy, the actual value of a Goodness Point can be made to vary. The value can go up or down according to the purchasing power or exchange rate.

Exchange rates of Goodness Points for Goodness Power change unpredictably randomly and constantly without the player's foreknowledge. A player never knows what reward she might earn for any one Good Deed. The game conveys that any individual Good Deed cannot be chosen merely for the reward it brings. Good Deeds are rewards in themselves. Sometimes exchange rates can even go negative and a player actually loses Goodness Points, having to sacrifice in order to perform a Good Deed. However, overall randomness is statistically biased so that a player usually comes out ahead. Players can adjust the game's level of difficulty by changing its randomness, range of random changes, and baseline Goodness Point value. This feature adds interest and luck to the game.

For example, a player may be randomly injured or handicapped through illness or accident. The value of his Goodness Points is adjusted to a lower baseline in proportion to his disability until he is healed, if possible. However, he may obtain help from other players or borrow from the Community Goodness Fund to survive. When a player is injured while committing a Misdeed, he also loses Stature, and the value of his Goodness Points is automatically depressed until she is forgiven or finds forgiveness, although still eligible to receive others' help.

The game also resets the baseline of a player's Goodness Point exchange rate or purchasing power in several predictable ways that add further incentives to practice good values. Several factors raise the baseline value of a player's Goodness Points in known amounts for a preset period of time and proportionally to the magnitude of the deed. In approximate increasing order, these include: storing Goodness Points in a Community Goodness Fund, a special fund that makes interest-free loans to needy others; initiating or inventing an original Good Deed; initiating Good Deeds on her own; helping another player; anonymously donating Goodness Points, deeds, or a needed organ to a special bank for the needy; maintaining good family relations; and teaching a son or daughter to do Good Deeds.

Direct awards of actual Goodness Points are rare. A player receives a gift of Goodness Points upon graduation from school. Also, family members of a disabled or injured avatar obtain a gift of Goodness Points for a limited period of time.

Reasons for lessening Goodness Point value baselines for preset periods of time include: Misdeeds, especially to family members, letting competition override decency; neglect of responsibilities to family, neighbors, and community; and excessive hoarding or accumulation of Goodness Power and Goodness Points without using them.

External factors can also affect the value of a player's Goodness Points: increase with good weather, victory in a war, a good economy and a crime-free society; or decrease with major random life setbacks, for example, illness, disaster, economic downturns, or war.

A player has wide latitude in shaping her avatar's unique character, path and in initiating action, rather than merely reacting. As the game progresses, players make many individual and group decisions. Initiating a Good Deed increases Goodness Point value.

LEVELS

Family support and life get high priority in the game. A beginning player always starts out in a family, his first level. His avatar carries the mark of his family. The new player must do Good Deeds for family members to start building Stature that allows advancement to the next level of community. Within a family, a gift of Goodness Points to a child or spouse is allowed and encouraged and counts as a Good Deed. However, mentoring a child counts more. The exchange rate increases for a parent who regularly teaches his child Good Deeds. A player can find shelter and temporary rest in a family after a severe setback, and start over. Family members of a player injured while doing a Good Deed all receive a compensation of actual Goodness Points for a period of time.

Levels are largely defined by the player's coherent life path that forms his identity and offers benefits to society. School is a player's first level of community experience. New opportunities for Good Deeds include community service and charity. In general, self-improvement is a lifelong undertaking that counts as Good Deeds. However, the main goal now is learning Mobujemoland's history, culture, heroes and ideals. A player takes courses and must pass tests. These provide valuable data about the player's thinking and learning. Successfully finishing a course counts as a Good Deed that earns a badge for that course and adds a measure of Stature. Male and female players have the same opportunities, although only females can give birth.

Educational material and tests are tailored for individual children according to parent-set priorities. The game can offer special needs players their own coursework or preventive or remedial classes. For example, a child with a weakness in reading is progressively taught reading skills to earn points.

To graduate, a player must choose two heroes from the game's mythology as role models. A commitment to heroes provides a path that requires a preset progression of Good Deeds that define the life's work, character, and reputation of the player's avatar. As he makes his way through life and completes each challenge on his path, he earns a badge. A player's Stature is increased as badges accumulate, much like the Scouts.

A player's Goodness Hero embodies ideals of goodness that include Spirituality (Moses, Jesus, Buddha, Mohammad), Beauty (Beethoven, Renoir, Michelangelo), Charity (Mother Theresa, Bill Gates), Knowledge (Einstein, Freud, Galileo), Politics (Lincoln, Gandhi, King, Mandela), Exploration (Columbus, Magellan, Salk), Athletics (Babe Ruth, Michael Jordan), and Freedom (Thomas Jefferson, Judah the Maccabee).

The second hero is a Work Hero, who embodies specific professions to benefit society and the common good, for example, artists, scientists, writers, teachers, doctors, and others. The player must place prominent markers of his heroes on his avatar.

Since the total goodness in Mobujemoland depends on the individual goodness of its Citizens, graduation from school is an important ceremony. The player must declare his life's goals in the tradition of his heroes and take an oath. His avatar is decorated with the heroes' marks, and he receives a large package of Goodness Power for use in the next level.

After graduation, a player always gets a better rate of exchange for Goodness Points earned from Good Deeds on the path that emulates her chosen heroes. He is thereby encouraged to remain on a consistent and sensible path toward his ideals and society's benefit.

Upon graduation, a player's avatar becomes a full Citizen. A Citizen is subject to the full range of rules, rewards, and obligations.

All Citizens who choose the same Goodness Hero and follow similar paths become members of a League together with that hero's other followers. Each League has a Mark, Meeting Hall, and fraternal rituals.

The number of Leagues expands regularly because players can request to include Goodness Heroes or subsets of professions or vocations. In their requests, they must study historical material about the composite hero and come up with an articulate justification. For instance, a new Broadcast Journalist League with Edward R. Morrow as the hero has recently formed within the Freedom League. A player must belong to one League and one League only. A player may switch heroes whenever he advances a level, but must return to school to learn about new heroes and at some cost of Goodness Points and Stature.

League members cooperate in many ways, but also compete with each other to advance ranks as Officers by excelling in Good Deeds along their chosen paths. For example, members of the Entertainment League compete to design entertaining games and quiz contests. Members of the Writers' League publish and critique their works. Such creative input is considered by the editors of the game and points and medals are sometimes awarded.

All Leagues have the same structures of levels that require advancement. League members compete to earn badges that increase their Stature. Stature is also increased by special recognition with prizes like the Pulitzer and Nobel. Players may contribute text relevant to the League's subject or submit suggestions and display original art, poetry, and writings.Steve's avatar is El Greco, after Greece's innovation of democracy. The avatar wears a hat similar to Abraham Lincoln's because his hero is Rooselinchill—a composite of Lincoln, the Roosevelts, and Churchill—one of many heroes in Mobujemoland's mythical history. He had been an important president eons ago. Together with other players who follow a similar path, Steve's avatar is a member of the Rooselinchill League. Nathan's avatar is also in the Freedom League, but is a broadcast journalist known as Pulrow (Pulitzer + Morrow). Nathan is a bit more advanced than Steve, having completed much of his course work and school challenge.

The game provides factual historical information in text or images, including holograms, and creates challenges based on historical events. Sometimes the challenge may be intellectual, and progress may depend on learning and passing quizzes, completing brain teasers, or fashioning winning strategies. For example, The Rooselinchill League (Their hero is Rooselinchill, a composite of Abraham Lincoln, Theodore Roosevelt, Franklin D. Roosevelt, and Winston Churchill.) is made up of all the players who chose a path of serving the community as political leaders. Members' avatars must wear tall black Lincoln hats as their mark. Challenges along the Rooselinchill path include hard labor of a woodsman, developing exceptional story-telling skills, educating oneself, practicing law, entering politics, becoming president and struggling with political challenges and the Civil War, freeing slaves, studying nature, coping with polio, coping with reconstruction, urbanization and industrial-labor crises, the New Deal, World War II, and many other challenges. Completing challenges along the path earn Goodness Points and badges. As a player progresses, Stature in the League grows and can progress along levels.

The Rooselinchill League's levels parallel Mobujemoland's governmental structures. Following graduation, a player's first level is the Neighborhood Civic Council. He must accumulate sufficient Stature and become its Chief Officer before advancing to the City Civic Council, the State Civic Council, the Regional Civic Council, and finally to the Grand Civic Council. The player must get elected, and along the way can also earn badges by holding civic and public service jobs like judge, police commissioner, and public health officer. As a politician, the player can also propose and work toward Good Deeds for the environment and the unfortunate. However, building hospitals and roads is also important. The Chief Officer of the Grand Civic Council is also the President of the Mobujemoland and is a member of the Civic Council of All Good Lands, together with presidents of other lands. Its only powers are to unite to do Good Deeds and adjudicate disputes on a universal scale. If human rights are violated by any government of any land, its Citizens lose Goodness Point value. Rogue Lands of game-generated Ungoods that do not subscribe to this aim are excluded,

The Civic Council of All Good Lands is akin to an alliance, and forms one side of the Universe Council, akin to our United Nations. The Universe Council can negotiate with the Band of Rogue Lands to maintain peace. Most of the time this is sufficient to keep things in line, but in rare circumstances, the Universe Council can declare war.

Whatever the League, every player is a Citizen and must also belong to a Civic Council (neighborhood, city, region, state, land). Stature in a League enhances the exchange rate of the common pool of Goodness Points of the Civic Council and adds to a player's reputation. By performing Good Deeds as a group, Civic Councils at each level compete with each other for Goodness Points that are partially distributed to each member. A Civic Council can raise Goodness Points to spend for its Good Deeds.

While being on the winning team earns Goodness Points, a player can also earn Goodness Points for good citizenship and contributions to the common community good, both inter- and intra-Civic Council cooperation, peacemaking and conflict resolution, problem solving, consensus building, and other leadership qualities. A player who does not support his Civic Council risks erosion of his Goodness Point value.

ETHICS AND PUNISHMENT

The Good Life is designed so that disputes, appearances of impropriety, conflicts of interest, crimes, and Misdeeds violating values are not unusual. These provide opportunities to learn morality and ethics. Both parties can go to the public Justice Room to resolve their dispute. If they resolve the dispute by themselves, they earn extra Goodness Points for problem solving and achieving a fair resolution. If they cannot do so, judges and juries made up of other players come into the room to attempt a fair solution.

Since each player is a member of a family and a League and is a Citizen, he may face ethical or moral conflicts and must choose one from among several Good Deeds, consequently neglecting the others. For example, the player may face a decision that calls for a choice among preserving the environment, attending to her family, or chasing an enemy. He may lose Goodness Point value for choosing one Good Deed and not choosing the others. While the program will offer an explanation, the player may appeal to a judge.

Enemy Ungoods are generated by the game itself as ethical and moral challenges to players. For instance, they can infiltrate corporations, schools, or the government of Mobujemoland in disguises. When discovered, chased and caught, they might initially continue clever deceptions or evasions until they answer questions and succeed in mental or physical challenges that only actual Citizens can complete. When uncovered, they are either exiled to their Rogue Land or transformed back to their true appearance and are jailed until they redeem themselves.

Punishment is always tempered with mercy and compassion. Any act of violence is immediately adjudicated by game masters and usually results in suspension or expulsion. Serious offenses are publicly adjudicated by avatars of game masters themselves in widely publicized trials, potentially adding to the Shame of a player.

An offender is given the benefit of the doubt and is innocent until proven otherwise. An offender may temper his punishments or even entirely redeem oneself by initiating on his own a substantial program of Good Deeds. Punishments always include bringing a public display of Shame via a mark on the avatar and contrition of a preset duration. Other punishments include banishment from the game altogether, prison sentences with opportunities for rehabilitation, community service without points, and loss of Goodness Power. Sentencing a Citizen can call on fine moral and ethical judgments and require deliberation and thoughtfulness, rather than black and white thinking.

After graduation, sufficient Stature enables the Citizen to become a defender—the higher his Stature, the more powerful the player in his defender role.

Mobujemoland is always facing one or more threats that require cooperation from every Citizen. Internal hazards can include a crime wave, poor city services, or just plain indifference to values and lack of community spirit. Civic Councils must cope with these threats. A campaign against criminals can be full of action.

A variety of external threats is posed by exotic morally and ethically corrupt Ungoods from Rogue Lands, each with different abilities and strengths, who frequently attack alone or in groups. Ungoods are generated and controlled by the game itself and not by players. Ungoods can foment internal trouble on a family, school, neighborhood, city, or state level, or threaten the entire land. They can infiltrate the society and form sleeper cells, only to later transform Citizens in a sudden terrorist attack. They can infiltrate corporations or government and defraud the public of Good Lands. Governmental Civic Councils on the appropriate level must fight fires, epidemics, environmental damage, crimes, natural disasters, or the threat of tyranny they cause. When the Civic Council of the Universe calls on Mobujemoland to defend world peace or human rights against a threat from a Rogue Land, it must comply.

One or more Ungoods or Rogue Land and can ally to form raiding parties or armies. The stronger the Ungoods, the greater the challenge, and the more Goodness Power a player and his Civic Councils need to cope with it. If an Ungood harms a friend and causes him a setback, a player can invest Goodness Power and in turn earn Goodness Points by helping him and his family. A player can also earn a badge or medal for courage.

The Ungoods damage, steal, and destroy property, but they do no explicit violence to player's avatars. Instead, they harm or injure Citizens by transforming them into Ungoods like themselves, in whole or part, or by recruiting them to spy. Medics can reverse such damage. While her Stature is increased, the injured avatar pays some Goodness Power.

How do the defenders eliminate threats without violence? First, they must repair the damage caused by the attacking Ungood and earn Goodness Points for cooperation and leadership. In case of a war, Mobujemoland's president attempts to make peace with Rogue lands in the Universe Civic Council, or at least make a deal for a cease-fire or a truce. A Good Land can encourage and support rebels in a Rogue Land to overthrow its malevolent government. When peace is achieved, all players' Goodness Points increase in value.

If these measures fail, defenders undertake a military action to capture and transform Ungood leaders and others, or if necessary, invade their lands or planets. Members of the Defenders League, marked by George Washington wigs, perform the Good Deeds of planning leading these efforts. In case of war, all Citizens' Goodness Point values drop markedly as a tax to energize weapons. Weapons vary in power and sophistication according to the Stature of the player's avatar. Tanks and other vehicles, planes, boats, submarines, and helicopters are common, and some battles take place in outer space. Some work in a laser-like fashion to immobilize hostiles, and others use nets and other devices to trap. A defensive campaign may be quite lengthy and elaborate and involve a whole variety of actions, chases, strategic planning, and even espionage. All players have an active role consistent with their skills, including drivers, medics, foot soldiers, and pilots. However, even during such a campaign, players must meet their other responsibilities to community and family.

Finally, an attacking Ungood is captured or an army defeated. Instead of destroying Ungoods violently, the community and individual players invest Goodness Power to forgive and dexterity to transform the enemy into an asset for family, neighborhood, city, state, or society. There are elements of chance built in that make this Good Deed less or more difficult, and robotic machinery requiring dexterity that makes the job easier or harder. Repairing and maintaining the machinery is also a League.

Transformations of Ungoods vary in intensity and helpfulness. The entire population of a Rogue Land can be transformed to a Good Land by a process of elections through the efforts of the Civic Council of Good Lands by pooling Goodness

Powers of each Citizen. In general, the more helpful the Ungood's final role, the more Goodness Power a player and his group earn back. For example, an Ungood can be turned into a harmless rock, requiring little investment and earning little rewards. Or it can be transformed into a specialized humanoid who serves society, requiring a great joint investment from many Civic Council members and returning a greater reward to each. Freeing humanoids and turning them into humans can earn a player an enormous increases in the value of their Goodness Points.

Defensive efforts cannot exceed an average of 20 percent of the time of the game, although there may be rare times of peak danger and demand for much more.

AFTER THE GAME

At the end, the game finally gives a five-minute warning so players can wind up unfinished tasks. Because the illusory power created by the interactive devices in Nathan's home media room is so great, transitioning out of The Good Life takes longer. The game brings the boys back to reality by reviewing illusions that appeared magical and teaches them how these were done technically. It quizzes them about this lesson, adding to their wealth. It also examines their state of mind and inquires about which images appear vivid, and again dissipates their intensity.

On the whole, both Nathan and Steve increase their Stature, but Steve's point wealth ends up very low that day, with much less potential Goodness Power available for the next session. The game records that there was much illusory action that day and alerts the boys' parents to possible residual effects. It also tracks each boy's neurocognitive developmental status along sensorimotor, Socialization, Values Education, Education Enrichment trajectories, and other measures. Today, Nathan's parents learn that his sign-reading skills are improving every day and his freedom from distractibility is increasing.

The game reports that Steve is much slower and less dexterous than usual. Upon seeing this information that evening, his parents wonder at the cause. Steve does have a cold and is taking medication; is his cold or the medication slowing him down? Or is he demoralized and partially giving up because his confidence is bruised by the major setback early in the session?

The next day, when Dick asked him, Steve sulked and whined that the game is unfair. Dick agreed, but reminded him of prior days when he was more successful, and of his nicely growing Stature. Dick spent much time also trying to convey to Steve the lesson that life is not necessarily always fair, and that sometimes we do get hurt when we try to do something nice for others. Dick asks, "Is that a reason to lose confidence and not do your best and give up? Aren't Good Deeds ultimately their own rewards?" He gives examples from his own life. Steve listens.

But Steve is not convinced. Steve then says that the game is boring, and he is not sure that playing The Good Life is all that it is cracked up to be; maybe he'll quit it and start another game. Dick and Jane discuss the situation. Together with his pediatrician, they have been suspecting for some time that Steve is vulnerable to depression, which runs in both Jane's and Dick's families. They have already learned the importance of helping Steve practice positive thinking and self-soothing at home. It is now time to move to the next step.

Today, for the first time, father and son access a therapy module recommended for Steve by the FITGOALS™ and the pediatrician. This module identifies negative thoughts and teaches positive thinking. From now on, they revisit this training module regularly, especially when Steve reacts to events in his life in an excessively demoralized manner. The Good Life receives input from this module and adjusts to subtly support Steve and buttress his self-esteem whenever it sags, while teaching his coping skills as well as proactive and optimistic attitudes, without giving him an unfair competitive edge.

These methods are a start, but Jane will eventually take Steve to a psychiatrist for more comprehensive evaluation and prevention of more serious depression. As part of this effort, Steve and his parents will later agree to enroll Steve anonymously in a special FITGOALS™ research module that advances science by evaluating different methods of preventing or delaying the onset of depression.

Index

A Space Odyssey, 155

abuse, 53, 105, 106, 107, 116, 164, 169

academic performance, 74

addiction, 7, 50, 112, 115, 120

adolescent, 11, 54, 76, 121, 122, 148, 175, 180, 196

advertising, 75, 78, 118, 124

aggression, 30, 33, 50, 65, 70, 76, 94, 106, 116, 161

Ainsworth, 13

American Academy of Pediatrics, 52, 65, 119

anxiety, 43, 56, 78, 160, 163, 174, 184

anxious, 19, 121, 163, 178

apologizing, 42, 104, 171, 200

apology, 99

appliance, 66, 113, 119, 133

arts and crafts, 25

assertiveness, 76, 93

attention, 1, 3, 13, 15, 18, 19, 22, 24, 25, 33, 43, 47, 51, 53, 54, 55, 56, 57, 58, 59, 60, 63, 64, 65, 66, 73, 75, 76, 77, 90, 96, 98, 107, 121, 122, 146, 166, 184, 193, 196, 197, 198

attentive, 27, 108, 170

autism, 71

avatar, 155, 185, 193, 194, 196, 199, 200, 202, 203, 204, 206, 207

away from the wall, 17, 128, 137

baby formula, 52, 53

baby video, 2, 15, 25, 52, 53, 55, 58

Barney, 64

beauty, 22, 73, 95, 112, 172, 200

bedroom, 10, 80, 131

bedtime, 178

Blade Runner, 155

block, 17, 39, 48

Blue's Clues, 63, 64

blueprint, 12, 13, 89

board game, 79

boctaae, 70

boredom, 19, 33, 54, 168

Bowlby, 13

brain, 8, 10, 11, 12, 13, 14, 15, 21, 22, 26, 28, 50, 51, 52, 54, 55, 56, 59, 60, 61, 62, 63, 65, 67, 73, 88, 89, 90, 91, 92, 105, 115, 146, 148, 150, 151, 153, 158, 159, 184, 193, 197, 204

breast-feeding, 52, 53

bullied, 43, 76, 93, 116

bully, 43, 75, 76, 80, 116, 139

choreographed, 21, 90

Christakis, 60, 123

collaboration, 1, 19, 27, 32, 68, 69, 70, 98, 140, 158, 159, 166, 168, 171, 180 collaborative, 5, 15, 27, 31, 67, 74, 96, 98, 99, 153, 160, 167, 170

communities, 14, 28, 31, 70, 76, 77, 117, 183, 189

community, 13, 22, 26, 28, 40, 51, 76, 77, 82, 93, 94, 95, 96, 108, 118, 121, 182, 185, 186, 190, 191, 193, 195, 197, 200, 202, 204, 205, 206, 207

competitiveness, 70, 109

computer literacy, 67, 80

conflict, 11, 17, 58, 83, 108, 109, 115, 122, 161, 170, 172, 173, 189, 195, 205

consensus, 41, 48, 98, 100, 101, 103, 205

considerate, 94

consideration, 29, 30, 42, 43, 66, 117

consumer, 28, 57, 66, 78, 119, 122, 123, 185, 186, 191

consumerism, 50, 65, 67, 78, 112, 124

consumers, 57, 75, 118, 122, 124, 149, 156, 184

courage, 91, 95, 96, 151, 207

creativity, 22, 31, 33, 40, 51, 58, 63, 71, 73, 151, 152, 159, 160, 161, 191, 195

critical thinking, 29, 32, 78

curiosity, 31, 33, 58, 62, 63, 73, 78, 142, 173, 175

curious, 58, 62, 63, 84, 175

dancing, 43, 73

day care, 34

decorate, 47, 79, 149

depressed, 19, 88, 121, 154, 174, 178, 201

depression, 43, 71, 78, 96, 121, 169, 170, 174, 198, 209

developmental trajectory, 21, 24, 34, 83

discipline, 31, 32, 57, 73, 94, 102, 104, 105, 106, 107, 112, 117

DNA, 13

do no harm, 6, 12

doctor, 30, 52, 75, 116, 158, 159, 160, 163, 164, 166, 167, 171, 179

Dora the Explorer, 64

Dragon Tales, 63, 64

draw, 28, 73, 79, 80, 90, 91, 151

drawing, 55, 63, 64, 69, 90

dreams, 56

earphones, 18, 29

ecology, 12, 55

educators, 53, 147, 197

emissions, 116

emotional health, 15, 31, 183

empathy, 22, 55, 69, 71, 89

Erickson, 13

esthetic, 162, 163, 170, 175

ethical, 18, 29, 30, 42, 47, 78, 88, 93, 94, 96, 118, 122, 156, 194, 196, 197, 199, 205, 206

ethics, 119, 156, 197, 205

etiquette, 29, 115, 117

eye contact, 17, 54, 169

eye-to-eye, 18

face-to-face, 18, 51, 56, 65, 66, 70, 71, 83, 110, 116, 119, 122, 137, 146, 149, 160, 171, 174

family life, 1, 2, 4, 5, 8, 10, 11, 14, 15, 20, 24, 25, 27, 42, 82, 92, 102, 108, 113, 117, 128, 134, 146, 166, 182, 183, 185, 186, 191

family time, 18

fear, 56, 62, 95, 96, 104, 113

filtering, 3, 5, 39, 42, 71, 86, 133, 185

FITGOALS, 144, 182, 183, 184, 185, 186, 188, 189, 193, 194, 196, 197, 198, 199, 209

food groups, 23, 26, 102

Frankenstein, 155

fraud, 75, 118, 154, 156

Freud, 13, 203

fully present, 12, 18, 19, 29, 30, 50, 53, 60, 64, 117, 121

furniture, 7, 16, 17, 38, 39, 182

gamers, 10, 191, 192

gender, 11, 28, 64, 68, 70, 78

Generation Digital, 115

genetic, 11, 51, 88, 89

Golden Rule, 29, 69, 88, 89, 90, 93, 94

Golem, 154

gratitude, 28, 30, 99, 159, 188

gtg, 70

guidelines, 15, 24, 34, 35, 36, 37, 43, 48, 52, 53, 65, 103, 109, 110, 111, 128, 129, 133, 137, 183, 184

hardwiring, 12

Harlow, 13

Harvey, 151

heroes, 28, 95, 190, 202, 203, 204

holding environment, 158, 166

homework, 10, 39, 47, 71, 74, 80, 83, 115, 117, 140

humanoid, 149, 152, 154, 208

identity, 11, 68, 70, 75, 76, 80, 113, 175, 180, 194, 200, 202

imagination, 19, 31, 33, 58, 63, 73, 152, 155, 160, 190, 191

imaginative, 73, 152

impulsive, 109, 116, 161

impulsivity, 11, 91, 161

infancy, 4, 11, 15, 21, 30, 32, 34, 57, 60, 88, 89, 92, 109, 148, 150

infant, 13, 25, 34, 55, 58, 89, 91, 148, 158

instant messaging, 28, 39, 70, 71, 83, 140, 163, 170, 171, 172, 174

intelligent toys, 3, 150, 153, 154, 183

Into the Minds of Babes, 53

intuition, 8

invent, 3, 33, 58, 150, 151, 172

invention, 151

inventiveness, 1, 63, 152, 161, 182, 191

Island of Dr. Moreau, 155

Island of Lost Souls, 155

junk food, 102, 115, 193

kindergarten, 62

kindness, 28, 30, 35, 55, 64, 66, 71, 92, 93, 100, 117, 167, 190

Kohut, 13

leader, 13, 14, 76, 98, 99, 101, 103, 104, 155

leadership, 5, 22, 28, 42, 63, 98, 104, 115, 128, 195, 200, 205, 207

learning how to learn, 26

librarian, 40, 74

lingo, 70, 120

lol, 70

Mahler, 13

mammals, 11, 190

marketing, 42, 48, 52, 75, 118, 123, 146, 194

media consumption, 4, 10, 23, 24

Media Plan Worksheet, 39, 41, 47, 48, 128, 129, 132, 133, 134, 137

menu, 23, 34, 39, 40, 68, 100, 129, 137, 184, 185, 188, 189

merchandising, 53, 115, 118, 124, 156

miracle, 11, 21

moral, 4, 11, 15, 18, 22, 26, 30, 38, 40, 41, 43, 46, 76, 91, 92, 93, 94, 96, 148, 152, 153, 156, 161, 192, 196, 199, 205, 206

motivated, 88, 109

motivation, 18, 19, 74, 103, 169

motor, 25, 32, 38, 43, 60, 64, 67, 73, 74

MP3, 2, 4, 29, 112, 120, 162, 163, 166, 167, 169, 170, 174, 183

music, 18, 21, 24, 27, 35, 54, 59, 74, 77, 84, 89, 119, 123, 162, 163, 166, 169, 170, 173, 175, 191

nature vs. nurture, 13

networking, 70, 116

neuroscience, 182

night terrors, 56, 62

nightmares, 56, 161

obesity, 20, 43, 102

older kids, 3, 27, 30, 32, 59, 70, 74, 81, 98, 110, 120, 133, 148, 171, 196

out of control, 1, 121

overstimulation, 116

Overuse Syndrome, 50, 115, 120

Pac Man, 191

parent nearby, 130, 136, 137, 139, 140

peek-a-boo, 56

physical arrangement, 17, 38, 39, 73, 137, 138

Piaget, 13

plasticity, 50

play materials, 148, 159, 163, 167

play therapy, 144, 158, 160, 161, 163, 164, 167, 168

politeness, 28

Pong, 191

portal, 41, 82, 113, 183, 184, 186, 189, 199

Portal Page, 42, 48, 134

preschooler, 4, 64, 71

privacy, 17, 29, 39, 71, 73, 80, 118, 124, 167, 184, 189, 194, 197

private, 17, 79, 112, 119, 131, 151, 171, 172, 185

productive, 10, 13, 73, 122, 146, 152, 190, 195, 199

psychotherapy, 158, 159, 169

punishment, 55, 69, 71, 92, 94, 104, 105, 115

Purple Rose of Cairo, 155

reading, 2, 10, 35, 64, 69, 73, 74, 79, 80, 82, 124, 131, 160, 165, 184, 197, 198, 203, 208

relational artifacts, 152, 153, 154, 155, 156

research, 5, 10, 15, 21, 41, 50, 53, 78, 118, 131, 146, 153, 156, 197, 209

respect, 27, 28, 30, 32, 35, 42, 52, 71, 74, 86, 89, 92, 93, 94, 100, 104, 107, 108, 131, 133, 161, 163, 167, 188, 196

restricting, 4, 23

robot, 155

Rutter, 13

savviness, 112, 115, 119

scaffolding, 55, 62, 158, 188

Schor, 78

screen time, 24, 27, 50, 54, 66

self-esteem, 6, 31, 46, 58, 69, 73, 74, 75, 76, 78, 90, 159, 175, 209

sensorimotor, 26, 40, 43, 46, 58, 60, 73, 164, 208

shopping channel, 59

shy, 71, 83, 174

shyness, 43, 71, 82

Simone, 155, 156

singing, 18, 53, 54, 63, 73

smartphones, 2, 23, 70, 123

solitary time, 83

spam, 79, 114, 124

spanking, 107

spatial arrangement, 16, 25

special needs, 6, 25, 31, 35, 40, 43, 46, 62, 184, 186, 193, 196, 203

spiritual, 3, 15, 22, 40, 51, 96, 152

Spitz, 13

Star Trek, 155, 194

Star Wars, 155

Stepford Wives, 155

stimulation, 12, 19, 28, 30, 33, 34, 43, 54, 55, 58, 59, 61, 63

Super Mario Brothers, 191

supervision, 17, 35, 56, 196

talents, 13, 31, 40, 51, 73, 159

teen, 94, 122

Teflon, 19, 114

television, 3, 10, 23, 50, 52, 65, 80, 123, 183

tell time, 80

temperament, 43, 51, 58, 109, 131, 132

terrible twos, 103

texting, 18, 29, 47, 70, 71, 77, 116, 139, 150

therapeutic space, 159, 161, 163, 164, 166, 167, 174

therapy, 1, 110, 154, 158, 159, 160, 161, 163, 164, 166, 167, 170, 174, 209

toddler, 54, 60, 91, 112, 188

transitional object, 62, 150, 152, 174

Tron, 155

Turkle, 152

vacant, 19

violence, 15, 53, 56, 62, 65, 66, 67, 70, 75, 76, 77, 81, 92, 93, 94, 95, 105, 115, 116, 117, 193, 195, 198, 206, 207

virtual reality, 31, 70, 191, 192, 194

Weekly Menu, 128, 129, 132, 133, 137, 140, 183

worried, 66, 83

younger kids, 4, 57, 110, 133

AUG 1 1 2010

GLENCOE PUBLIC LIBRARY
3 1121 00338 6338

RENEW ONLINE AT
www.glencoepubliclibrary.org
select "My Library Account"
OR CALL 847-835-5056

DATE DUE

NOV 1 5 2010

DEMCO, INC. 38-2931